/ / ᶜ⸗ᵗₛ

C000007185

The
Straw
Manual

*A practical guide
to cost-effective straw
utilization and disposal*

The Straw Manual

*A practical guide
to cost-effective straw
utilization and disposal*

Bill Butterworth

LONDON
E. & F. N. Spon
NEW YORK

First published in 1985 by E. & F. N. Spon Ltd
11 New Fetter Lane, London EC4P 4EE
Published in the USA by E. & F. N. Spon

© 1985 Butterworth

Printed in Great Britain by J. W. Arrowsmith Ltd, Bristol

ISBN 0 419 13660 6

This paperback edition is sold subject to the condition that it shall not, by way of trade or otherwise, be lent, resold, hired out, or otherwise circulated without the publisher's prior consent in any form of binding or cover other than that in which it is published and without a similar condition including this condition being imposed on the subsequent purchaser.

All rights reserved. No part of this book may be reprinted or reproduced, or utilized in any form or by any electronic, mechanical or other means, now known or hereafter invented, including photocopying and recording, or in any information storage and retrieval system, without permission in writing from the publisher.

British Library Cataloguing in Publication Data

Butterworth, Bill
The straw manual: a practical guide to cost-effective straw utilization and disposal
1. Straw
I. Title
633.1   SB189.3
ISBN 0-419-13660-6

Contents

Preface

This book is not about what might be done with straw one day. It is about what can be done with it right now on the farm. It is about *today's* technology, allowing decisions to be made about what to do with straw *this* year in order to minimize costs, minimize management problems and maximize yields and profits.

It can be read cover to cover, chapter by chapter or dipped into as a reference book. Whichever way, this manual is aimed at helping the farmer to understand the technology and do the job that has to be done – profitably.

Action Plans which summarize the relevant data and ease decision-making appear at the end of each chapter. The final part is a Decision Planner which will help with tackling the job in hand, on the day.

Acknowledgments

Grateful thanks are given to many individuals and organizations for their assistance, support and provision of information. Special acknowledgment is made to the fine work done by ADAS and their staff who made efforts to provide a great deal of up to date data for this publication.

Thanks for assistance are also given to ICI Plant Protection Division and to John Deere who, through their annual Award to agricultural writers, sparked this project off by giving the award to the author in 1982.

'I will never put my name on a piece of writing that does not have in it the best that is in me.'

Bill Butterworth
1 January 1985

1
Straw production

STRAW AS A RAW MATERIAL

The growth in production of cereal straw in the UK has far outstripped an increase in the use of straw. Current surpluses are estimated at something in excess of 7 million tonnes per year. The total production is about 12.5 million tonnes leaving a fairly constant use on the farm of about 5 million tonnes. The quantity of excess production is expected to continue to rise rapidly to 10 million tonnes by 1990. Straw production is, of course, mainly in the arable areas where there is less livestock enterprise to utilize the straw in the traditional way (see Figs 1.1, 1.2 and 1.3).

Most of the wheat straw is burned because it is of comparatively low feed value and it is mainly grown in areas of low livestock enterprise (Table 1.1).

Table 1.1 Estimated quantities of straw baled in England and Wales (1983)

Cereal	Quantity baled (million tonnes)	(%)
Wheat	3.12	41.2
Barley	4.26	56.2
Oats	0.20	2.6
Total	7.58	100.0

Source: MAFF, 1984. *Straw disposal and utilisation – a review of knowledge.*

What is not generally realized is that straw is a very highly variable material with its value as a material very dependent on species, variety, climatic conditions, soil type and husbandry. These factors result in significant differences in physical, chemical and biological characteristics.

It is important to see what straw is made up of in order to see its potential use (Table 1.2).

Table 1.2 Typical cellulose, hemicellulose and lignin content of grass and straw

Forage	Cellulose (%DM)	Hemicellulose (%DM)	Lignin (%DM)
Fresh grass	25	28	4
Straw	41	29	11

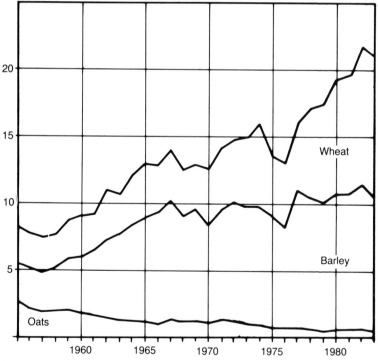

Fig. 1.1 Cereal production in the UK between 1955 and 1983 (million tonnes).
(Source: **MAFF Agricultural Statistics).**

Straw surplus (1000 tonnes)

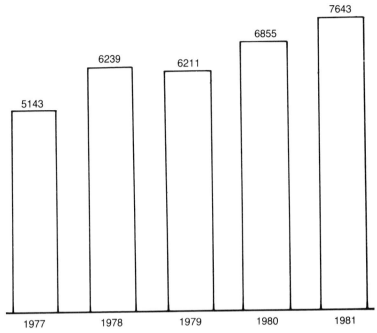

Fig. 1.2 **Surplus straw in England and Wales 1977-1981.**

These carbohydrate materials have potential use as food, fuel or a substitute for an organic chemicals industry. The problem is that the material has a very high bulk, is difficult to handle economically and the chemical value of the material is not easy to mobilize and actually use in practice. A further problem to all these possible uses is the highly variable nature of the product. It is commonly thought that straw is relatively uniform but important chemical and physical differences do occur. Table 1.3 shows, by way of example, that straws are highly variable in feed value. Wheat, for example, is generally regarded as inferior to barley and never very high. The table, however, shows the range of values that may occur in practice.

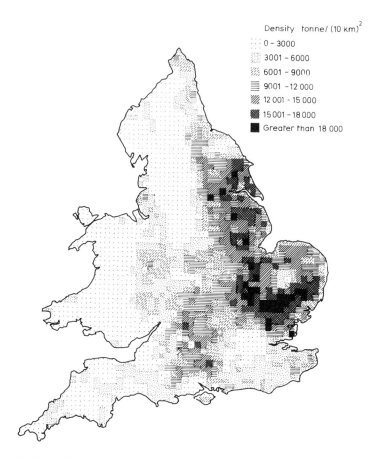

Density tonne/ (10 km)²

0 - 3000

3001 - 6000

6001 - 9000

9001 - 12 000

12 001 - 15 000

15 001 - 18 000

Greater than 18 000

Fig. 1.3 Distribution of cereal straw production in England and Wales.

Table 1.3 *In vivo* DOMD and ME values of straw

Cereal	'n' samples	DOMD (%)		ME (MJ/kg DM)	
		Mean	Range	Mean	Range
Wheat	34	42	30-53	6.5	3.9-10.5
Barley	38	45	33-56	6.8	3.4- 9.4
Oats	7	50	38-59	7.9	5.2-10.1

POSSIBLE UTILIZATION

Feed

Clearly, from Table 1.3, the material can be used as an animal feed. This subject will be gone into in some detail in Chapter 5. There is, however, a clear limit on how much straw could be used as animal feed.

Table 1.4 suggests the MAFF maximum amounts of untreated and treated straw which could be eaten by various animals and is used to estimate limits to straw consumption as animal feed.

Table 1.4 Maximum straw consumption by ruminants

	Maximum straw consumption (kg/day)	
Cattle and calves	*Untreated*	*Treated*
Dairy cows and heifers	4.0	6.0
Beef cows and heifers	6.0	9.0
Other cattle, 2 years old and over	4.0	6.0
Other cattle, 1 year old and under 2	3.5	5.0
Calves, 6-12 months	2.5	4.0
Calves, under 6 months	1.0	1.5

Using the 1981 Census data, Steve Larkin of Silsoe College has calculated that the cattle population of 7.9 million could consume 5.7 million tonnes of treated cereal straw. If fed untreated, this would reduce to 3.8 million tonnes. It could displace significant areas of grassland currently supplying the hay and silage eaten by these animals, if straw is transported from areas of surplus to areas of deficit.

On a within county basis, the major cereal growing counties could feed only a small proportion of the straw they produce because of insufficient animal numbers. The following counties would be able to use less than 20% of their production: Bedfordshire, Hertfordshire, Cambridgeshire, Essex, Lincolnshire, Norfolk and Suffolk. For this reason and also the rate at which we could adjust husbandry to change, it is unlikely that even with a major effort, we will actually use more than a million extra tonnes of straw per year in the next five years. There is, then, no foreseeable likelihood of this outlet being able to mop up all the available surplus, or keep up with the envisaged expansion.

Plate 1.1 As the grain crop grows, so does the straw crop.

Fuel

We already have machinery to burn straw in a baled form as a fuel to provide heat for domestic use, livestock houses, grain drying or other agricultural use. This will be covered in detail in Chapter 6.

Straw is already used, as straw in bales, in quite large quantities in this way. This is, and will remain, an attractive method of utilization because no capital expenditure nor running costs are involved in further processing. However, the idea of conversion to 'bio-gas', largely methane, is attractive because gas is storable, pipeable and controllable.

Methane production from wet 'slurry' has its problems in Northern Europe because the process needs to be kept warm. However, a relatively 'dry' process can still work. Cornell University have been carrying out a research project on 'dry-fermentation'. The dry-fermentation term is actually a misnomer, says William J. Jewell, professor of agricultural engineering, who headed the Cornell project. In the context of the Cornell work, which was funded by the US

Energy Department, the term means the absence of drainable water. For example, the raw material might have a moisture content of only about 70%, while in conventional fermentation slurries the water content is approximately 90%.

The difference in moisture content is important, emphasizes the researcher Wolfgang Baader at the Institute for Agricultural Research (Braunschweig, West Germany). That's because some of the bio-gas – in liquid fermentation, as much as 30% – is consumed in maintaining the reaction at 95°F, the optimal temperature for bacterial metabolism. The less water in the slurry, the less energy is needed to keep it warm.

A successful scale-up from a 175 ft^3, reactor to a 3900 ft^3 vessel, Jewell says, strongly suggests that the process is close to economic viability. In Jewell's present process, 1 tonne of animal manure is used as the bacterial inoculum for every 4–5 tonnes of straw. Tight control of moisture content was found to eliminate the need for buffering with sodium bicarbonate, which was originally used to prevent the formation of acids that poisoned the bacteria. Even in the cold winter months, the researchers reported, only 5% of the energy output was needed to maintain the reaction temperature. The problems encountered with these plants so far is that the capital expenditure involved is high, they don't work without problems and the economic value of the gas produced in practice is not what the experimental procedure suggests is possible.

The immediate future, as a fuel, then appears to be with straw *as straw*, either in bales or, less likely, compressed possibly as briquettes. 'Compressed' bales and briquettes will be discussed in Chapter 4, and straw as a fuel in Chapter 6.

Straw as a chemical source

Potentially, because straw contains the elements carbon, hydrogen and oxygen, it could be used as the basis for a complete organic chemicals industry; as a substitute for oil. Straw can be used to make furfural from with many organic chemicals can be made. Plants, each to use a quarter of a million tonnes a year, have been examined for building in the UK. None have yet got very far in the planning process. It does not seem likely that such plans will progress at this time and, even if they did, it would take five years to get a plant on stream. In the immediate future this is not a likely home for our surplus straw.

Straw as a building material

Various industrial processes have been developed for turning straw
into a variety of boards at a range of densities from a rigid insulation
material to a dense board similar to chipboard. This outlet has
received a fillip recently with the introduction by Marshall of
Gainsborough of a process to make a dense board. This market,
however, has so far seen its fluctuations and has never swept into a
prominent share of the market. While the indications are that this
share will grow, this outlet is, like some of the others, unlikely to
swallow up that 7.5 million (and growing) tonnes of surplus.

INCORPORATION

That leaves incorporation as a means of disposal and utilization. It is
likely that this will be the most rapidly growing method of disposal in
the mid-1980s. It is this area that has caused so much interest with
the possibility of significant extra costs, management problems and
yield losses (see Table 1.5).

Straw, stubble and cereal roots consist mainly of lignin, cellulose
and hemicellulose. In soil about half the cellulosic compounds
decompose within six weeks but the lignin is relatively resistant to
decay. The breakdown is brought about by naturally occurring
bacteria and fungi which are present in the soil. There are a wide
range of micro-organisms which attack the straw and many of them
have been identified. Work is currently in progress which may lead to
ways of enhancing their performance and speeding the breakdown of
the straw by inoculation.

The AFRC (Agricultural and Food Research Council) has been
involved in very extensive research for some years. It is already
known that the state of the soil will affect the way in which the
breakdown proceeds. Under waterlogged (anaerobic) conditions,
fermentative bacteria produce acetic acid which is toxic to emerging
seedlings. Other bacteria can enter the root and depress growth. If the
soil is too wet the straw degrading soil micro-organisms can compete
with the seed for oxygen and nutrients and delay germination or
emergence.

Studies at Letcombe Laboratory and the GCRI have shown that
under certain conditions cellulolytic fungi and nitrogen fixing bacteria
can interact and accelerate straw degradation.

Table 1.5 Straw disposal 1984

		Straw baled and removed		Straw ploughed in or cultivated		Straw burned in the field (a)		Total (b)	
		1983	1984	1983	1984	1983	1984	1983	1984
Wheat	Hectares ('000)	651.0	638.3	34.5	183.8	958.4	1040.8	1643.9	1862.9
	per cent	39.6%	34.3%	2.1%	9.9%	38.3%	55.8%	100.0%	100.0%
Winter barley	Hectares ('000)	672.9	731.6	5.9	32.9	158.2	151.7	837.0	961.1
	per cent	80.4%	79.8%	0.7%	3.6%	18.9%	16.6%	100.0%	100.0%
Spring barley	Hectares ('000)	647.1	491.2	17.8	24.4	142.0	59.7	806.9	575.3
	per cent	80.2%	85.4%	2.2%	4.2%	17.6%	10.4%	100.0%	100.0%
Oats	Hectares ('000)	67.8	66.5	0.4	1.9	11.8	9.2	80.0	77.5
	per cent	84.7%	85.7%	0.5%	2.5%	14.8%	11.8%	100.0%	100.0%
Total	Hectares ('000)	2038.8	1927.6	58.6	243.0	1270.4	1261.4	3367.8	3431.9
	per cent	60.5%	55.9%	1.8%	7.1%	37.7%	37.0%	100.0%	100.0%
Number of holdings	Thousand	64.7	65.8	3.1	7.8	24.6	25.4	0.6(c)	0.1(c)

Notes 1: The results of the Straw Survey which was conducted as part of the 1984 Survey of Cereals on Farms. Details of straw harvested, ploughed in or burned were obtained from a controlled sample of 2400 cereal growing farms and the results were grossed up to give estimates for England and Wales as a whole. The figures are subject to sampling errors.

2: The straw areas shown have been estimated by applying the percentages derived from the survey to June census final crop areas.

Key: (a) Excluding area of stubble burned after removal of straw.

(b) Total crop areas used are final figures from the June censuses and exclude minor holdings.

(c) The numbers of holdings disposing of straw in different ways add up to more than this total because some holdings dispose of straw in more than one way.

Plate 1.2 The straw crop is not what it used to be.

Much of the work done by ADAS has been first rate and will be reviewed in Chapter 5.

Being positive about straw incorporation, straw has a value as a fertilizer and will make a useful contribution to the physical properties of the soil as organic matter. Moisture retention, ease of working and biochemical activity will all show at least marginal improvements. The fertilizer value of wheat straw, for example, is about 3.4 kg of nitrogen, 3.0 kg of phosphate and 12 kg of potash per tonne. A heavy, 9 tonne per ha crop of wheat straw would therefore supply 30 kg of nitrogen, 27 kg of phosphate and 108 kg of potash per hectare. Not a lot, but of some value.

Burning

At the time of writing this introduction in January 1985, straw burning had not been banned or severely curtailed. It is, however, likely that increasing constraining regulations will be brought into force. Despite that possibility, burning remains, and is likely to remain, an option. It will, however, be increasingly difficult to carry out and will become less attractive as the process of burning requires more and more resources and provides an increasingly difficult managerial input into the process itself and the management of fields which have only their centres actually burned. This subject will be discussed in detail in Chapter 2.

ACTION PLAN — OVERALL

1. Draw up a 'Straw Balance' of expected production of each variety and of expected use requirements.
2. Detail the exact type of straw available and expected quantities.
3. Detail exact utilization requirements.
4. Detail the farming system and examine the possibilites for:
   (a) use as bedding or feed (see Chapter 5);
   (b) use as a fuel (see Chapter 6);
   (c) off-farm sale;
   (d) harvesting methods and costs (see Chapter 4);
   (e) incorporation – detail the fields and methods to be used (see Chapter 3);
   (f) examine the possibilities for burning.
5. Read the relevant Chapters. Make decisions. Draw up a management plan.
6. Implement the plan, modify according to circumstances on the day. Record results.

DECISION PLANNER — OVERALL

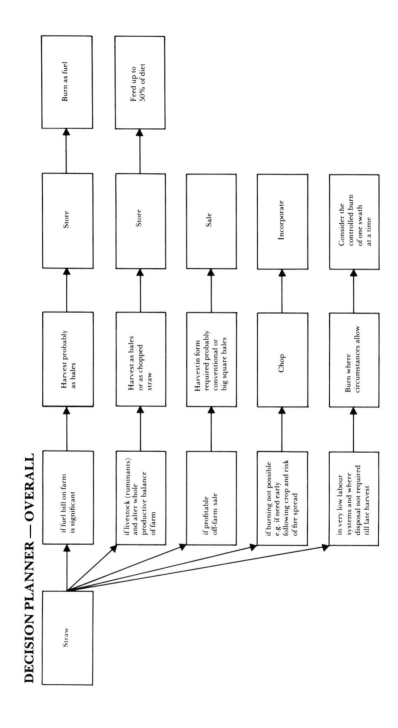

Straw	if fuel bill on farm is significant	Harvest probably as bales	Store → Burn as fuel
	if livestock (ruminants) and alter whole productive balance of farm	Harvest as bales or as chopped straw	Store → Feed up to 50% of diet
	if profitable off-farm sale	Harvest in form required probably conventional or big square bales	Sale
	if burning not possible e.g. if need early following crop and risk of fire spread	Chop	Incorporate
	in very low labour systems and where disposal not required till late harvest	Burn where circumstances allow	Consider the controlled burn of one swath at a time

2
Field burning

METHODS OF BURNING

Consequences of burning

There are many methods of burning straw. Some are more effective than others, some are easier to control, some have a greater devastation of wildlife. Burning from a line upwind, for example, does tend to allow wildlife to escape, burns rapidly and effectively but easily damages downwind hedgerows. Possibly the most effective burn is obtained by a method sometimes called 'vortex burning' after the saturation bombing of World War II. Lighting the outside of the field in a ring produces a draught which sucks the fire rapidly into the centre. The hedges are not seriously affected, the burn is very rapid and thorough but wildlife trapped within the circle is devastated. It is worth noting that some desiccants under some conditions can make straw burn more fiercely but, conversely, they can also reduce smoke and help create a more even, clean burn.

The fact is that common sense and the law will not allow burning except within a narrow band of possibilites and this is likely to be restricted progressively in future.

The 1984 NFU Straw Burning Code set out the following details which will certainly be updated by the NFU and legislation.

1. Never burn where there is a likelihood of causing damage or annoyance to the public.
2. Never burn during weekends or bank holidays.
3. Never burn when the weather is unsuitable.
4. Never light a fire before sunrise or later than 1 h before sunset.
5. Never burn without at least 2 people to supervise each fire.

6. Always create a firebreak of at least 15 m for hedgerows, trees or telegraph poles and 25 m for other features at risk.
7. Always clear straw from beneath overhead powerlines and above-ground electricity, oil or gas installations.
8. Always burn straw against the wind if at all possible.
9. Always inform your neighbours and, if required, the fire brigade and district council of your plans.
10. Always incorporate ashes within 36 h.
11. Always use the checklist when preparing to burn.

Know the law – including your local byelaws.

Code of Practice
BURN ONLY . . .
if there is no alternative: Straw can be used as bedding, a feedstuff or fuel and there is a growing industrial market. You may also be able to incorporate it into the soil;
on weekdays: Do not burn on Saturdays, Sundays or bank holidays;
in daylight: Do not light any fire before sunrise or later than 1 h before sunset (sunset falls 30 minutes before lighting up time);
in suitable weather: Extra care is needed if burning is necessary after a prolonged spell of hot, dry weather. Remember you must incorporate the ashes afterwards;
in low winds which assist combustion: Do not light a fire when the wind exceeds force 3 (8–12 mph – a wind of force 4 moves small branches) or the direction of the wind is likely to create a hazard or annoyance from smoke or smuts, especially near houses, airfields, public buildings and horticultural enterprises;
well away from public roads: Do not burn within 100 m of a motorway or dual carriageway.
BEFORE BURNING . . .
spread the straw: either during or after combining. Spreading produces a more complete burn and can reduce the risk of smoke and smut problems. Chopping may produce a further improvement (remember that some desiccants make straw and stubble burn more fiercely);
check your local weather forecast: prepare to burn only if forecast conditions are suitable;
inform your neighbours (including the managers of publicly-owned land) the county fire control and the district council environmental health department where required by local byelaws;
divide the area to be burned into blocks of no more than 10 ha: each block

should be bounded by a firebreak prepared as in Table 2.1. To comply with the byelaws you must not allow two blocks to burn simultaneously if they are 150 m or less apart, even if the other fire is on another farm;
prepare firebreaks as in Table 2.1.

Table 2.1

Feature	Straw clearance	Additional action
10 ha block to be burned	clear a 5-m strip	cultivate or plough entire strip
Hedgerows Trees Telegraph poles	clear a 15-m strip and move straw at least 25 m from the feature	cultivate entire strip or plough 5 metres of it
Residential buildings, Thatched buildings, Buildings, structures, or machinery made of glass or combustible material, Stacks of hay or straw, Accumulations of combustible material, Standing crops, Woodland and nature reserves, Scheduled monuments made of combustible material	clear 25-m strip	cultivate entire strip or plough 5 metres of it
Other ancient monuments, e.g. standing stones	clear 3-m strip	
Non-combustible electricity, oil or gas installations	clear 15-m strip	

have readily available a stock of at least 5 fire beating implements and a crop sprayer or similar mobile container(s) with suitable hose attachment containing at least 500 litres of water for each block to be burned.

arrange for two responsible persons to supervise each fire. One, who must be experienced in controlled burning, should be placed in charge of the operation. The Agricultural Training Board runs courses on straw burning.

ensure you know where help can be found and summoned quickly in an emergency.

safeguard electricity, oil and gas installations. Clear straw from at least 15 m around poles, pylons and other above-ground installations.

take special precautions where overhead powerlines cross the field. Intense heat

Plate 2.1 Straw burning has much to commend it, but it will be progressively restricted.

and smoke can cause lines to flashover to the ground like lightning. Clear the straw well away from overhead lines.

burn straw against the wind if possible. If you have to burn with the wind, remove or burn at least 30 m of straw inside the fire break at the downwind end of the area.

WHILE BURNING . . .

keep children well away from the field.

watch the speed and direction of the wind and be prepared to stop burning if conditions change.

AFTER BURNING . . .

check that no straw remains alight. Return later to make doubly sure. Always make sure the fire is extinguished by nightfall.

incorporate all ashes into the soil as soon as possible, at the latest within 36 h of beginning to burn. Ashes left on the field are picked up in high winds and often deposited where they create a serious nuisance.

Legal controls
Many of the Code's provisions are backed by law, principally by the

byelaws. The following is a summary of the model byelaws published in 1984.

A ban on burning at weekends, bank holidays and at night (night begins 1 h before sunset and ends at sunrise).

Firebreaks of 25 m next to residential buildings, thatched buildings, glasshouses, scheduled monuments, buildings, structures, plant or machinery, the greater part of which are constructed of combustible material (other than straw removed in the construction of a firebreak). Either 5 m of the firebreak should be ploughed or the entire strip cultivated.

Firebreaks of 15 m (with a requirement to remove straw from that strip to a point at least 25 m away from the object in question) next to hedgerows, trees and telegraph poles. Either 5 m of the firebreak should be ploughed or the entire strip cultivated.

A limit of 10 ha on the area to be burned. Each 10 ha block must be bounded by a 5-m strip which has been cleared of straw and then ploughed or cultivated. Blocks must be at least 150 m apart when burned simultaneously.

Supervision of two responsible persons, of whom one must be in charge of the operation and experienced in the burning of straw or stubble. Firefighting equipment must be readily available. The model byelaws stipulate at least 500 litres of water in one or more mobile containers 'together with a means of dispensing the water for firefighting purposes', and a stock of at least five fire beating implements.

Ashes must be incorporated into the soil within 36 h of the start of burning.

Where byelaws are adopted by a district council they have the force of law. Each contravention can incur a fine of up to £2000.

Check your local byelaws to find out whether they differ from the model.

Other laws with a bearing on straw burning are as follows:

Clean Air Act 1956 provides that action may be taken against anyone who allows smoke (or smuts emitted in smoke) to endganger public health or cause a nuisance.

Heath and Safety at Work etc. Act 1974 provides an integrated system of law dealing with the health and safety of virtually all people at work, and protection for members of the public where they may be affected by the activities of people at work.

All persons engaged in straw or stubble burning whether employers, self employed or employees, are covered by the Act. If straw or stubble burning places anyone at risk prosecution may result. The maximum fine on summary conviction under the Act is £2000. There is no limit to the fine if conviction is on indictment. Compliance with this Code will assist farmers and others in meeting their legal responsibilities.

Highways Act 1980 makes it an offence to start a fire within 50 ft of the centre of a road so as to inconvenience passers-by or damage the highway.

Public utility installations are given special protection by the following statutes: *Public Health Act 1936, Water Act 1945, Gas Acts 1948 and 1972, Pipelines Act 1962 and Land Powers (Defence) Act 1958.*

Checklist
Every time you burn straw or stubble check before you burn:

- the weather forecast
- your neighbours know of your plans
- you are fully insured
- the county fire control and district council have been informed if required
- your firebreaks are correct
- five hundred litres of water and five firebeaters are on site
- where help can be found in an emergency
- the block to be burned does not exceed 10 ha
- each block is at least 150 m from any other being burned

Remember you may not start a fire before sunrise or later than 1 h before sunset.

BYELAWS

The byelaws in your area may be different from other areas nationally or even over the hedge. The model national byelaws are laid out below but do check the local situation with the local authority.

Model byelaws for straw and stubble burning – Home Office Circular 1984. *Reproduced by kind permission of HMSO, these byelaws are Crown Copyright.*

Extent of byelaws

1. These byelaws shall extend to ....

Restrictions on burning

2. No person shall, on agricultural land, commence to burn any straw or stubble remaining on such land after the harvesting of any cereal crop which has been grown thereon, or cause or permit to commence the burning of such straw or stubble at any time:

(a) during the period beginning 1 h before sunset and ending at sunrise;
(b) on any Saturday, Sunday or bank holiday.

3. No person shall commence to burn or cause or permit to commence the burning of such straw or stubble unless the area in which it is intended to burn such straw or stubble is more than 150 m from any other area where such straw or stubble is being burned.

4. (i) No person shall commence to burn or cause or permit to commence the burning of any area of such straw or stubble unless that area does not exceed 10 ha and:

(a) without prejudice to sub-paragraphs (b) and (c) below, is bounded on all sides by a firebreak constructed by removing so far as is reasonably practicable all such straw from a strip of land not less than 5 m width and either cultivating or ploughing that strip of land;
(b) subject to sub-paragraph (c) below, where any part of that area is within 15 m of any of the following objects, that is to say any hedgerow, tree or telegraph pole, a firebreak constructed by removing so far as is reasonably practicable, and to a distance of not less that 25 m from that object, all such straw from a strip of land not less than 15 m in width between that area and that object and either cultivating that strip or ploughing not less that 5 m in width of that strip;
(c) where any part of that area is within 25 m of any of the objects specified in paragraph (ii) below, a firebreak is constructed by removing so far as is reasonably practicable all such straw from a strip of land not less than 25 m in width between that area and that object and either cultivating that strip or ploughing not less that 5 m in width of that strip.

(ii) The objects referred to in paragraph (i) (c) above are:

(a) any residential building;

(b) any structure having a thatched roof;

(c) any building, structure, fixed plant or machinery the greater part of which is constructed of combustible material or glass or both;

(d) any scheduled monument the greater part of which is constructed of combustible material;

(e) any stack of hay or straw;

(f) any accumulation of combustible material other than straw removed in the construction of a firebreak;

(g) any standing cereal, oil seed or pulse crop;

(h) any woodland or nature reserve.

(iii) Where for the purposes of constructing a firebreak required by this byelaw it is necessary to measure a distance from a tree the distance shall be measured from the trunk of the tree.

5. No person shall commence to burn or cause or permit to commence the burning of such straw or stubble on any day unless not less than one hour's notice has been given on that day to the County Fire Control of the .... County Fire Brigade (and, if available, such officer of the .... District/Borough Council as the Council may appoint for the purpose of receiving such notice).

6. No person shall burn or cause or permit the burning of any area of such straw or stubble unless during the whole time the material is burning the operation is under the supervision of at least two responsible persons present at the burning of that area, of whom one is in charge of the operation and is experienced in the burning of straw or stubble.

7. No person shall, without reasonable excuse, burn or cause or permit the burning of any area of straw or stubble unless during the whole of the time the material is burning the following means for fighting fire are available at the burning of that area, that is to say:

(a) not less that 500 litres of water in one or more mobile containers together with a means of dispensing the water for firefighting purposes; and

(b) not less than five implements suitable for use for firebeating purposes.

8. The occupier of the land on which such straw or stubble has been burned shall not, without reasonable excuse, permit any ash or carbonized residues, not incorporated into the soil of the land, to remain for a period of more than 36 h after the commencement of the burning on an area on which straw or stubble has been burned.

Defence
9. In proceedings against any person for an offence under byelaw 3 or 6 above it shall be a defence for that person to prove that he had taken all reasonable precautions and exercised all due diligence to avoid the commission of the offence.

Penalty
10. Any person contravening any of these byelaws shall be liable on summary conviction to a fine not exceeding £1000.

Interpretation
11. In these byelaws:
'combustible material' means material capable of undergoing combustion;
'combustion' means combustion by oxidation with the production of heat, usually with incandescence or flame or both;
'nature reserve' has the same meaning as in section 15 of the National Parks and Access to the Countryside Act 1949; and
'scheduled monument' has the same meaning as in section 1 of the Ancient Monuments and Archaeological Areas Act 1979.

Revocations
12. The byelaws made ....
....
.... are hereby revoked.

Developments

It is already the case that, in most instances, £2000 is the maximum fine. It is also true that a number of loopholes are likely to be closed. For example, the byelaws relate to cereal straw; rape falls outside these rules.

It may be that significant progress is made down the restriction road by the time this book is published. One halfway option is to burn one strip of straw at a time. This would significantly reduce all problems associated with straw burning including smuts travelling far downwind.

SWATH BURNING – AN ALTERNATIVE

The diagrams in Fig. 2.1 show a system of swath burning which could solve many of the objections to straw burning. The tractor required for a system based on a 4.5 m (15 ft) combine table would require a 3 m (10 ft) cultivator on the back, a side rake to push a swath at least 7 ft sideways to the left and a boom with a gas flame mounted to give 3 m (10 ft) reach on the right. The flame boom needs to be purpose-built with fail safe features built in.

Pass 1 (into the page) pushes the swath to the right of the diagram and cultivates strip 1 enough to act as a fire break.

Pass 2 (towards the reader) comes back down the other side of the next swath to push two swaths close to each other and cultivates a fire break on strip 2.

Pass 3 repeats pass 1 with respect to the next pair of swaths *and* sets light to swaths 1 and 2 which are now isolated.

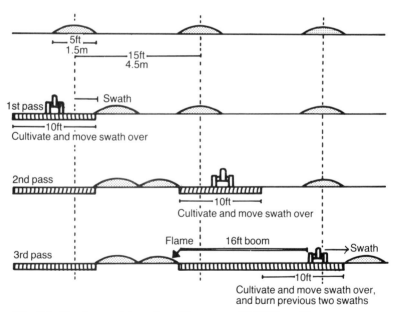

Fig. 2.1 Single swath burning. (Courtesy of J. Hildyard.)

Pass 4 is carried out later and incorporates the ash in the burned strip.

Headlands have to be cleared first in compliance with the byelaws and in order to isolate the swaths for a controlled burn. The system leaves the whole field treated more or less the same. It will also reduce the updraught and smut problem.

BURNING – EFFECTS ON SOIL FAUNA

The early experience with direct drilling showed quite clearly that residues of straw could lead to dramatic increases in slug population. It was not until 1979, however, that Edwards and Lofty (1979) of the Rothamsted Experimental Station published more detailed results of research into the effects of straw burning on a wider range of small animals in the soil and on its surface.

There are two effects of burning. Firstly, there is a significant and damaging rise in soil surface temperature during the burn. This actually kills off animals living on, at or near the soil surface. Secondly, organic matter on the surface is reduced and this, in the long term, affects populations of 'interested' fauna such as earthworms. These effects can be either beneficial or harmful. The burn does kill many pests such as aphids, but it also has a deleterious effect on soil structure when it reduces earthworm populations. The long term effects of earthworms on soil structure, aeration and drainage may have been underestimated.

Figure 2.2 shows the effects of different straw burning methods on a range of invertebrates that live on or near the soil surface. Clearly, a good burn does have significant effects in all the species involved with sampling with a suction sampler working at the surface.

In fact, as Fig. 2.3 shows, the burn did affect the population not only at the surface (where it was most marked) but also to the limits of the depth measures at 15 cm. The mites and Collembola (springtails) in Fig. 2.3 tend to live near the surface in any case, so it is the surface figures that are most affected.

As Fig. 2.4 shows, the larger invertebrates are also affected. Spiders (Araneida), thrips (Thysonoptera) and other insects were all shown to decrease by this research work. However, as Fig. 2.5 shows, there are cases of increases in populations following a burn. Enchytraied worms are such a case. Thrips (Thysonoptera) were almost eradicated by the burn.

The effect on aphids was interesting and important with baling decreasing populations by 31%, burning in rows by 90% and spreading and burning by 92%.

Deep burrowing earthworms were dramatically affected by three years of burning, but some shallow working species actually increased (Fig. 2.5).

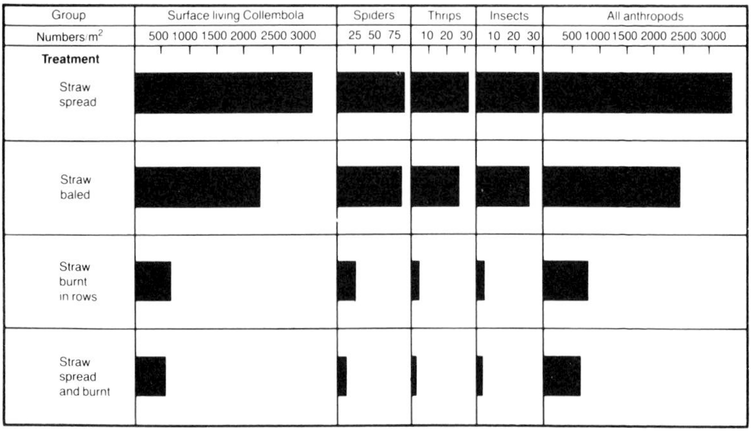

Fig. 2.2 **Numbers of surface living arthropods 10 months after straw removal (D-vac suction samples).**

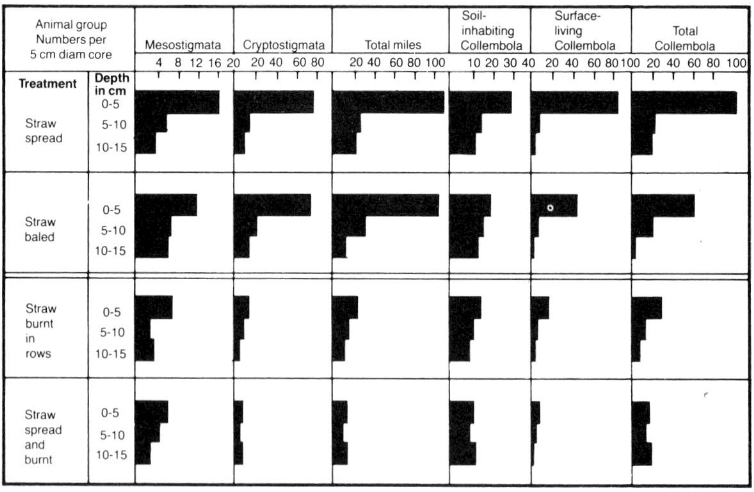

Fig. 2.3 **Vertical distribution of soil micro-arthropods 3 months after straw removal (soil samples).**

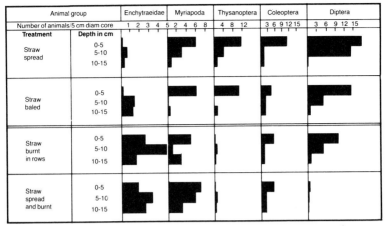

Fig. 2.4 Vertical distribution of larger invertebrates 3 months after straw removal (soil samples).

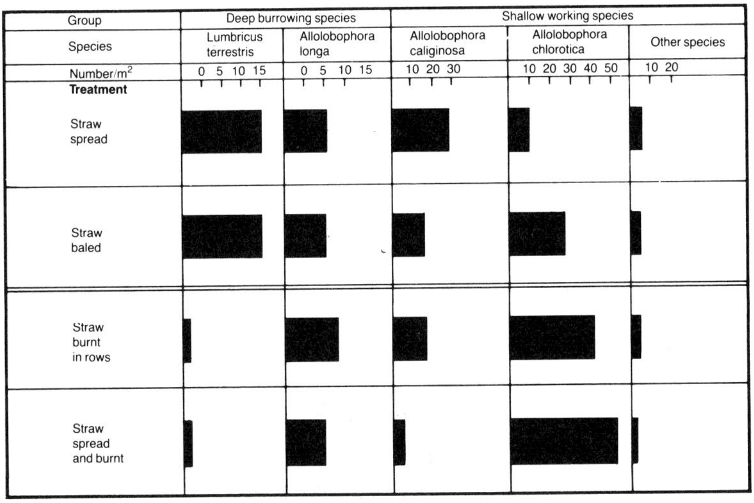

Fig. 2.5 Numbers of earthworms 4 months after the third successive year of straw removal.

The deep-burrowing *Lumbricus terrestris* will move significant quantities of straw from the surface to depths well below normal cultivation depth. In considering possible effects of burning on soil structure, the major effect on soil structure is most likely due to the deep burrowing earthworms and the best measure of their effect is usually considered to be their weight (biomass) rather than numbers. Figure 2.6 shows that burning can significantly reduce this biomass.

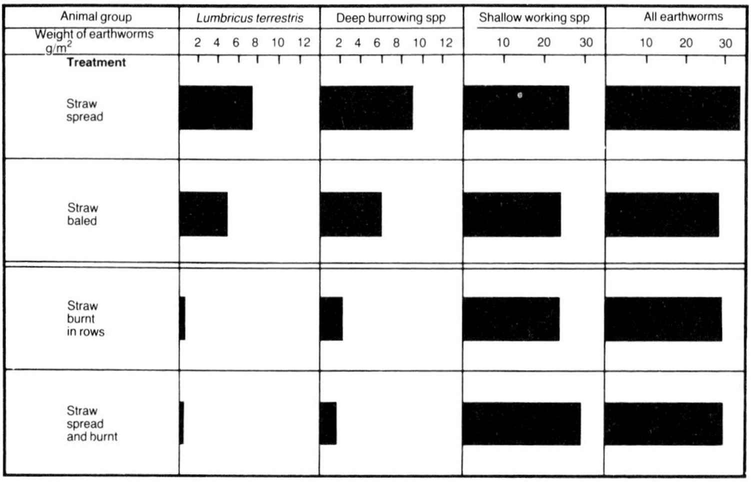

Animal group	Lumbricus terrestris						Deep burrowing spp						Shallow working spp			All earthworms		
Weight of earthworms g/m²	2	4	6	8	10	12	2	4	6	8	10	12	10	20	30	10	20	30
Treatment																		
Straw spread																		
Straw baled																		
Straw burnt in rows																		
Straw spread and burnt																		

Fig. 2.6 Biomass of earthworms 4 months after the third successive year of straw removal.

Incorporated straw

Partly because of inadequate burns, it is generally accepted that earthworm populations increase under direct drilling systems. However, the effects of burning on the deep burrowing species is potentially significant. The converse is also true; Fig. 2.6 also shows that straw spread and retained on the field results in increase in worm populations.

Pests

Generally, straw retention will tend to increase populations of surface living pests, such as slugs, thrips and aphids.

BURNING – EFFECTS ON FLORA

It is commonly said that a good burn is worth perhaps twenty pounds per acre. Certainly there are effects on weed seed survival and germination.

Blackgrass

Burning of straw can bring about considerable kill of weed seeds. In WRO experiments, on average around 30% of wild oats seeds and

50% of seeds of blackgrass and barren brome have been killed. Burning has also been shown to reduce dormancy in some seeds such as wild oats and generally produces conditions favourable to blackgrass and barren brome seedling emergence. These factors result in an increased proportion of autumn seedling establishment from the surviving seeds of these weeds. It is obviously of most value, if seedlings can be destroyed before planting the next crop. So burning will affect planning of subsequent cultivations and both pre- and post-emergence spraying.

In fact, soil surface conditions significantly alter the way the burn affects seed survival and germination. Straw burning does destroy many seeds, depending on the nature of the burn. It also encourages germination and establishment (Table 2.2).

Table 2.2 Blackgrass populations. Interaction of cultivation system and straw disposal method on numbers of viable *A. myosuroides* seeds per m^2 in the top 3 cm soil after drilling in November (Data derived from Moss, 1979)

Cultivation	Straw baled	Straw burnt	Decrease by burning (%)
Plough	347	231	33
Tine	1669	700	58
Direct drill	3372	1016	70

Straw burning destroyed many *Alopecurus myosuroides* (blackgrass) seeds lying on the soil surface and, compared with straw baling resulted in a smaller weed infestation in a subsequent direct-drilled winter wheat crop. The larger the amount of straw burnt, the higher the temperature reached on the soil surface and consequently the greater the control of the weed. Seeds that were fully imbibed were less susceptible to destruction by heat than were dry seeds. A shallow soil covering protected most seeds although some were stimulated to germinate. After straw burning the soil surface was a better environment for seed germination or seedling development than that under stubble. Differences in seedling numbers between treatments were not always proportional to differences in the numbers of viable seeds in the soil.

(*Source: Weed Research*, 1980, Vol. 20)

The position was summarised by S. Moss (WRO) in Fig. 2.7 which shows that effective straw burning can reduce infestations compared to where straw is baled and also slow down the increase in infestation

Blackgrass
seedlings/m²

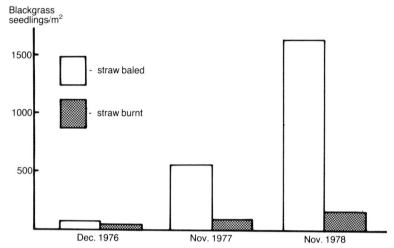

Fig. 2.7 Effect of straw disposal on blackgrass infestations over a 3 year period in winter wheat established under a time cultivation system.

where reduced cultivation systems are used. On land where straw is baled, population increases of over ninefold per year can occur. Straw burning on its own does not prevent a gradual build-up of blackgrass and consequently additional control measures are required.

Requirements for blackgrass control
A population model for blackgrass has been developed from which it is possible to determine the percentage weed control required to maintain a static weed population (Table 2.3).

Table 2.3 The annual percentage kill needed to maintain a static population

	Straw burnt	*Straw baled*
Ploughed	50	65
Direct drilled	88	92

These estimates show that there is a requirement for much better weed control in minimum cultivation systems, especially when straw is not burnt.

Brome

R. J. Froud-Williams of the WRO published a paper in March 1983 on the effect of the burn and various cultivations on *Bromus sterilis* (in *Ann. appl.Biol.*, 1983).

An initial population of 12635 seeds/m^2 declined by 85% between July and late August in uncultivated stubble without straw burning, but only 44% of seeds gave rise to seedlings. A further 10% had produced seedlings by late December and another 5% emerged between February and April. By April no viable seed remained.

Straw burning destroyed 97% of the ungerminated seeds on the soil surface and reduced seedling numbers by 94%. However, those seedlings which did survive formed many tillers and produced considerable quantities of seed.

Shallow cultivation reduced the seed population by 34% and in April there were 47% fewer seedlings on these plots than on direct drilled ones. Ploughing to a depth of 20 cm eradicated the weed; and although buried seeds germinated, they failed to emerge.

Seedlings emerging in spring failed to flower before harvest. Detailed laboratory investigation showed that although *B.sterilis* does not have a pronounced requirement for vernalization, chilling did accelerate flowering, while long-day photoperiods were required for panicle extension. Only germinated seeds responded to the vernalization stimulus.

Mr Froud-Williams' conclusions were: 'In the absence of a suitable selective herbicide, periodic use of the mouldboard plough remains the most realistic means of controlling this weed. Although this will not be acceptable to farmers totally committed to minimal tillage practices, useful suppression or containment may be achieved from a combination of straw burning and shallow cultivation, especially if crop rotation were also adopted'.

Wild oats

A paper published in *Weed Research* in 1975 by Wilson and Cussans of WRO put wild oats in context.

'An infestation of *Avena fatua* in two successive spring barley crops was monitored from March 1972 to June 1973. The numbers of viable seeds on the uncultivated stubble, viable seeds in the soil, and seedlings emerging in autumn and spring were assessed. Both seed and seedling populations declined when seeding was prevented in 1972, but increased in all situations where seeding occurred. The greatest increase occurred when the unburnt stubble was cultivated after harvest; the increase was less following straw burning. The greatest losses of newly shed seeds occurred when the stubbles were left uncultivated throughout the autumn, and delaying cultivation was more effective than burning in limiting the rate of increase.

Old seeds, already in the soil before 1972, were less dormant and produced a higher percentage of seedlings in 1973 than did seeds shed in 1972. Once in the soil, the greatest losses of seeds occurred during the spring, and it is likely that many seeds germinated but failed to establish as seedlings. It is suggested that the seed reserves of *A.fatua* in the soil are less persistent than previously reported, especially where mouldboard ploughing is replaced by tine cultivation. The periodicity of seedling emergence in spring was unaffected by the time or type of cultivation, by the age of the seeds or by burning the straw in the previous autumn'.

BURNING – ASH ABSORPTION OF HERBICIDES

It is now known that repeated annual straw burnings can, and often do, raise ash levels to a point where the free carbon absorbs sufficient of any soil applied herbicide to significantly inhibit its performance. This effect will also apply to high levels of organic matter from incorporation of straw.

It appears that although burnt straw residues may comprise only a small proportion of the total soil carbon, they may have a disproportionately greater influence on the absorptive properties of the soil.

The effect of burnt straw residues is likely to be greatest where they are retained close to the soil surface, as occurs with direct drilling. Tine cultivations will disperse residues to a degree dependent partly on the depth of cultivation. The greater the dispersion, the greater will be the dilution factor and, consequently, the less will be the likely effect on soil-acting herbicide performance. It follows that pre-crop emergence weed control becomes more important because early emerged weeds are more difficult to kill later on.

Says Steve Moss: 'Where adsorptive surface soil results in inadequate weed control by soil-acting herbicides, the following solutions are recommended:

1. If possible, apply soil-acting herbicides such as chlortoluron post- rather then pre-emergence. However, weed control may still be inadequate in highly adsorptive soils.
2. Use foliage-acting herbicides. Diclofop-methyl applied at 1.13

kg/ha in the autumn has given excellent control (greater than 95% of blackgrass on direct drilled sites where many soil-acting herbicides have performed poorly.

3. Rotational ploughing, once every 4-5 years, should bury the adsorptive surface soil layer and provide an environment in which soil-acting herbicides are likely to be more effective. This would also have the benefit of burying weed seeds to a depth from which few seedlings could emerge.'

ACTION PLAN — FIELD BURNING

1. Study the codes and byelaws.
2. Decide procedure and allocate men/machines.
3. Avoiding local complaints depends on:

    (a) complying with codes of practice, byelaws and regulations;
    (b) informing locals of the plan;
    (c) considering wind.

4. Consider the effect of the burning plan on:

    (a) the cultivation of the field;
    (b) the spray programme.

The burn (notes)

1. Burning up behind the combine, especially early in harvest, involves risk to the crops.
2. Resources are involved in a legal and successful burn.
3. Firebreaks are different from the rest of the field which is burned. Subsequent treatment may have to be different.
4. Disposal by the controlled burn (one swath at a time) can reduce risks and leave the whole field in the same condition.

DECISION PLANNER — FIELD BURNING

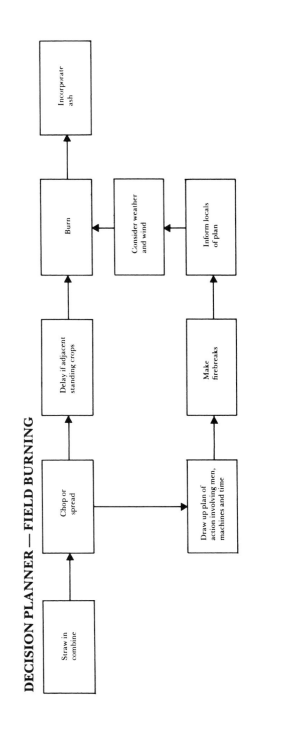

3

Incorporation of ash, stubbles and straw

THE NATIONAL SCENE

There is no doubt that historical evidence shows that burning straw off does have a very significant effect on the following crop. This effect will be seen in weed population, disease and in yield. The Letcombe Laboratory has long been advocating the burn. Table 3.1 shows the results of trials at Letcombe in 1977–8 and indicates a yield depression from incorporation. It should be remembered that this was one trial in one year only with very shallow incorporation to only 7 cm.

Table 3.1 Letcombe 1977–78

	Wild oats (%)	Wheat (t/ha)
Burn		
D. drill	0	8.51
Tine	0	8.14
Chop		
Surface	14	3.72
Disc	7	5.37
Rotavate	14	4.45

It has to be remembered that a total physical ban on straw burning in the field is, in practical terms, not feasible. This would involve the banning of bonfires in general. Nor, at the time of writing, has the Minister of Agriculture gone down that road very far. Burning is legal even if under fairly well controlled conditions. There is also a fair percentage of cereals that are baled, thus leaving a stubble. So incorporation of a full crop of straw is not the only outcome of the combine operation.

The national scene in 1983 and 1984 was shown up in MAFF survey statistics (Table 1.5).

Ash

The 'Revision of Model Byelaws for Straw and Stubble Burning' dated 20 March 1984 stated that:

'The occupier of the land on which such straw or stubble has been burned shall not, without reasonable excuse, permit any ash or carbonised residues, not incorporated into the soil of the land, to remain for a period of more than 36 hours after the commencement of the burning on an area on which straw or stubble has been burned.'

There is, then a commitment to incorporate ash although 'incorporate' is defined only in terms of not allowing ash 'to remain'. Some attempt has to be made to avoid the blowing of ash. This can be achieved with very light cultivation with harrows or almost any sort of discs.

This does not mean that direct drilling in the absolute sense is not possible where a burn is involved although it could, technically, be carried out into stubble. As far as a burn is concerned, direct drilling must be carried out into a scratched seedbed.

Stubble

There is no doubt that the amount of straw *utilized* rather than *disposed of* will gradually increase. As the pressure on burning increases and research allows straw to substitute for other increasingly expensive resources, straw will be an increasingly useful product. It will, however, always be a high bulk material of relatively low value. The

Plate 3.1 The plough is likely to be very important in straw incorporation in the next few years. Note the trash boards.

conclusion must be that, on well over half the cereal acreage, it will be stubble that has to be dealt with.

Incorporation

The area of straw ploughed in or incorporated in some way jumped dramatically in the autumn of 1984. The social and political pressure is bound to further increase the area. The practice of incorporation is, therefore, of increasing interest particularly in the cereal growing areas. Many farms who have tried incorporation do find advantages in simpler organization, more uniform treatment of fields and fewer local complaints. However, the fear of yield loss is of major concern and more important in the minds of many farmers than increased resources required to incorporate. Clearly, the effect on final margin is a matter of concern.

Plate 3.2 Shallow ploughing has much to recommend it. On many soils 15 cm (6 in) may be adequate.

EFFECTS ON YIELD

The evidence of the effects of incorporation on yield are variable. There is some evidence that on some soils, particularly light ones, there may be a very small effect to increase yields. Most of the available data on the majority of situations, particularly heavy soils, point to the possibility of loss of yield ranging from 0 to 10% depending on the methods used and the circumstances.

Table 3.2 shows the results of Experimental Husbandry Farms for the year harvested in 1984. Each EHF took the results of plots burned as an index of 100 for that farm. Clearly, on some farms chopping and ploughing to 20 cm (8 in) gave slightly higher yields but there were also reductions on other farms. All the EHF's carried out the 'core

Table 3.2 Results of straw incorporation trials 1984 (ADAS). Yields in tonnes per ha.

Site	Fenton, Lincs.	Tetsworth, Oxon.	Drayton EHF, Warks.	Bovingdon Hall, Essex*	Otley, Suffolk*	Boxworth EHF, Cambs.	Terington EHF, Norfolk	High Mowthorpe EHF, Yorks.	Bridgets EHF, Hants
Soil		Denchworth clay vale	Lower lias clay	Chalky boulder clay	Chalky boulder clay	Chalky boulder clay	Silty loam	Silty clay loam over chalk	Chalky silty loam over chalk
Variety	Flanders		Norman		Longbow	Longbow	Norman	Brigand	
Date drilled	15 Sept.		4 Oct.		24 Oct.	11 Oct.	4 Oct.	5 Oct.	
Nitrogen (kg/ha)						240	270†	190	200
1. Straw Burnt									
Direct drill	7.04	9.38	8.37	9.71	–	11.25	11.14	6.87	9.19 / 8.84
Various cultivation systems (with straw burned = 100** for each vertical column)	107	103	96	99	10.68 = 100	96	98	102	99
2. Straw chopped									
Plough 20 cm									
Shallow tines	99	–	99	94	96	90	96	96	96
Medium tines	94	107	100	93	90	92	93	96	95
3. Extra treatments									
Straw chopped:									
Incorporation + plough	110	97				96			99
Glencoe Soil Saver						93	95		
Rotadigger						95	95		
Tillaerator								96	
Dynadrive									
Tillage train						96			103
Tines, discs 5 cm					97	90			96
Tines, discs 20 cm						91			94
4. Direct drill into stubble		105			98				

* Not corrected to 15% MC.
† Other rates and timings of nitrogen tested.
** Minimum cultivation except at High Mowthorpe and Bridget where ploughed.

treatments' with other farms, particularly Boxworth carrying out a range of other possibilities.

The situation as described by these results, in what was probably a good incorporation year, is that marginal changes in yield are likely after incorporation. The question is how to limit yield reduction and produce a trend to yield increases rather than reductions.

The Plough

From these results the plough systems have an attraction in yield if not rate of work and energy requirement. An interesting result is that at the Tetsworth site where direct drilling into stubble produced a yield of 105. So direct drilling as such can be practised and it can work very well in stubble.

Experience in Germany

In Germany, they manage easily with straw incorporation on their clays and worry about the long term effects of incorporation on light land. The Germans pour in several times the power we are used to and that gives three problems; doing all the work on time, compaction, and the cost.

Most of the German soils are relatively light by our standards. Certainly those in Schleswig Holstein, the mecca of wheat growers, are so. Most of the farmers there still base their operations on the mouldboard plough and will have five to ten following passes in order to create a seedbed. Ploughing around 30 cm (12 in) deep is normal.

There are two alternatives in German thinking; to put the straw well under the seed or well over the seed. Most farmers incorporate into the surface and then plough deep leaving the seedbed way above the straw/soil layer. The idea over there would be to incorporate into the top 10 cm using chisels and discs and then plough to 30 cm (12 in). That mixes the straw and then buries it to leave a clean seedbed for conventional coulter drills.

On light land, where deep ploughing is still the rule, many German farmers and researchers believe that the straw builds up in the soil to make it too fluffy and advise a burn (which is still legally possible) or the sale of the straw crop off the land every 3–5 years.

There are farmers using the Horsch system which cultivates the top 5 cm only, throwing a mixture of soil and chopped straw up into the air. The idea is that the soil settles first in a layer over the seed and the straw on top. The system is basically a seed/fertilizer trailer in front of

Plate 3.3 Normal underbody clearances are quite adequate for dealing with burnt stubbles or even quite large amounts of chopped straw.

an Accord distributor mounted over a 4 m Howard rotavator. The seed is distributed from a cross-bar behind the rotor at ground level. The system can and does produce good crops on suitable soils which are non-smearing, level and not liable to distortion from wheelmarks.

Many UK farmers could follow the incorporate and plough system with, probably the equipment they already have. There is a problem. The inputs of time and diesel are unacceptable.

The indications from the best UK research are that we can do this job satisfactorily with lower inputs.

We have good research indicators, if not conclusions, from Letcombe, NIAE and ADAS. This suggests that incorporation to about 15 cm (6 in) will be adequate to spread the straw through sufficient soil to solve the technical problems (discussed below). If this were done and the soil not ploughed, oxidation would probably avoid the 'puffiness' problem met in Germany. Deep ploughing probably preserves the straw which would undergo relatively rapid oxidation if left near the surface.

Would this work? The trials at the NIAE suggest it would. Some trial work from Germany also suggests that this is where the answer lies. The figures are in the table but read them carefully and within the context of the UK research. This may produce a conclusion not reached by the German workers.

It is clear from Table 3.3 that the fields with the chisel cultivator only came out significantly worse in yield. On farm one, for example, the figures were 5.901 t/ha for the chiselled crop compared with 6.295 for the best.

That was, however, for only one pass with the chisel. There are all sorts of factors that would depress crop yield in such a regime. The British approach would be to leave the straw in the surface, allow a suitable delay and drill with a suitable drill.

Studying the table will allow a number of conclusions to be drawn provided it is remembered that the chisel was, with these trials, once over only.

UK conclusions from German experience
Incorporation followed by ploughing demands fine chopping with a target of at least 75% of the material chopped to less than 4 cm (1.5 in). Chopping of the stubble is desirable and a good even spread essential. Incorporation with two or three shallow passes, involving one or two passes of a straight legged cultivator and one of discs would do well. Each pass should be followed by an integral roller to get good soil/straw contact. This work would be to 5 cm, maybe a little deeper. Ploughing would then follow set practice in the UK probably to only 20 cm (8 in).

The alternative involving chisels and avoiding ploughing would be more profitable and faster, provided we learn to manage it properly and get good yields. The basic rule of a good chop, evenly spread is vital. Breaking the soil would be best done with a straight legged cultivator to make the soil 'boil' and get a good mix.

Experience in Denmark
The Danish experience is closer to UK conditions than those in Germany and the fact is that the Danes are tending to push yields up with incorporation although they accept some difficulty with heavy soils. The key to heavy soils in the UK may well be to leave the straw near enough to the surface for air to be active in its breakdown. Further, it is the mouldboard plough that looks like the safest option.

Figure 3.1 shows the results of Danish research from 1967 to 1982,

Table 3.3 Yield records of winter barley (following winter wheat). Commercial farm trials in Schleswig Holstein, 1983.

Farm	Treatment	Yield (t/ha)	Heads/m²	Head wt (g)	1000 grain wt (g)	Grains per ear	Volunteer wheat population m²
Rothenstein	Cultivator 1 pass only	5.901	1134.6	0.52	51.6	10.2	24.2
	Plough only	5.854	1094.2	0.54	52.4	10.5	3.1
	Shallow plough and deep plough	6.280	1145.6	0.55	52.3	10.4	0.1
	Cultivator and deep plough	6.295	1094.0	0.58	53.5	10.9	0.7
	Significant difference at 5% level	0.335	122.0	0.06	0.1	1.0	3.2
Brodersdorf	Cultivator 1 pass only	6.145	877.5	0.70	49.8	14.1	83.5
	Plough only	7.709	926.6	0.83	48.6	17.1	11.5
	Shallow plough and deep plough	7.557	878.5	0.77	48.0	16.0	16.3
	Cultivator and deep plough	8.112	673.0	1.21	44.2	27.5	6.4
	Significant difference at 5% level	0.562	54.9	0.06	1.3	1.5	20.2
Datgen	Cultivator 1 pass only	5.824	764.6	0.76	49.5	15.4	86.3
	Plough only	6.646	960.9	0.69	44.5	15.5	5.4
	Shallow plough and deep plough	6.492	954.0	0.68	44.7	15.2	31.0
	Cultivator and deep plough	6.330	861.3	0.74	48.5	15.1	6.4
	Significant difference at 5% level	0.720	79.8	0.05	1.6	0.7	26.2
Kitzeberg H$_7$	Cultivator 1 pass only	4.244	415.2	1.02	54.7	18.6	
	Plough only	4.545	396.9	1.15	54.2	21.2	
	Shallow plough and deep plough	4.886	432.3	1.13	52.2	21.6	
	Significant difference at 5% level	0.709	56.2	0.18	3.1	2.7	
Kitzeberg H$_8$	Cultivator 1 pass only	4.324	411.3	1.05	54.0	19.4	
	Plough only	5.076	454.7	1.12	52.3	21.5	
	Shallow plough and deep plough	5.294	487.5	1.09	51.6	21.1	
	Significant difference at 5% level	0.499	65.6	0.14	3.3	2.0	

Plate 3.4 Slatted mouldboards are unlikely to significantly reduce draught under most conditions, but may scour more easily under some soil conditions.

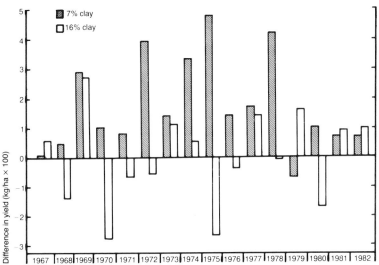

Fig. 3.1 Differences in yield of spring barley with straw incorporated by ploughing to 20 cm (8 in) compared with burning straw as zero. Years 1970, 1975 and 1980 were wet years (Figures from the Danish State Research Station at Ronhave).

i.e. 15 years of incorporating straw into the surface and ploughing 20 cm deep (only about 8 in). In nearly every year the yield on the lighter soil (shaded columns) was increased on the straw incorporation plots. The centre line is taken as the non-incorporated yield which averaged 4 t/ha in 1982. Yield on the light soils was increased in more years than not. On the heavier soils, the depressions would probably be associated with wetter autumns. In fact, statistical analysis indicates that the results are not highly significant but there are clear trends, yields can go up but there may well be problems on heavier soils. Note that the incorporation procedure was to harrow two or three times and then to plough with trashboards and not too deep.

UK experience

The Letcombe Laboratory has been fairly strong in pointing out the problems with straw and crop yield. Research and commentators there say that winter cereals do not compensate properly for the problems created by incorporated straw and yield is reduced. On average the loss of yield was 10% in 17 experiment years (ranging from +8 to −47%) after incorporating chopped straw with shallow tillage (5 – 7 cm), compared with burning followed by similar tillage (7, 15, 16). Even in the presence of stubble only, the yield loss averaged 5% (ranging from +3 to −17%) with shallow tillage. Thus, concluded Bob Cannell, previously of Letcombe, simplified or shallow tillage systems in the presence of relatively heavy yields of straw are not practicable.

However, the results from the Bosworth and Bridgets EHF's in the 1984 harvest indicate that it is possible to go quite some way down the road of shallow cultivation (see below).

Toxins

There has been much talk, possibly too much, about toxins from straw. There is good evidence that these can be a serious problem affecting seed germination at establishment.

Work at Letcombe has shown that when straw decomposes in wet anaerobic conditions substances can be formed that can be toxic to germinating seedlings. Of the substances formed, acetic acid can occur in sufficient concentrations to prevent or retard germination and to restrict the rate of elongation of roots and shoots. These effects, clearly shown in laboratory conditions, parallel the effects in the field although acetic acid has never actually been found under field conditions. The Northern Region ADAS put out the following notes in the autumn of 1983.

Straw breakdown in soil

The breakdown of straw in soil is a microbial process. Fertile soil contains large numbers, and many different types, of microbes ready to use plant material as food and to break it down. It has been estimated that there are 1 billion bacteria, 1000m of fungi and 1000s of tiny soil animals and algae in a thimbleful of fertile soil. This is equivalent to 1 tonne of microbes/hectare of soil!

The actual chemical construction of straw is complex but it consists primarily of:

	Approximate %	*Breakdown rate*
Water soluble materials (e.g. sugar)	8.12	days
Cellulose (a polymer of glucose)	38 – 44	months
Hemicellulose (a polymer of glucose and other sugars)	32 – 36	months
Lignin (a polymer of phenols)	10 – 15	years

As indicated above the rate of breakdown of the different chemical compounds varies widely. The water soluble materials are most readily broken down but are also constantly being formed as breakdown products of cellulose and hemicellulose which take a few months to degrade. Lignin however is only very slowly broken down over a matter of years but its residues in soil cause no problems. Some of the cellulose is chemically bound to lignin and therefore its rate of breakdown is reduced. The constituents also occur in differing amounts in the different parts of the straw, e.g. the nodes, and leaves contain more water soluble materials than the internodes.

Despite this chemical complexity, given the right conditions, in particular moisture, warmth and air, the microbes and enzymes naturally present in soil can break down straw without problems.

Factors affecting microbial activity are as follows:

1. Moisture – if too dry the microbes will not grow and multiply. If too wet air will be excluded (anaerobic).
2. Warmth – the warmer it is the faster the microbes work.

Plate 3.5 Crawler tractors may be important with incorporation of heavy wet soils.

3. Air – if conditions become anaerobic different kinds of microbes will take over. Some have been shown to produce chemicals which are toxic to seedlings (phytotoxins). The greatest risk period is the first two months depending on conditions after incorporation, since high concentrations of phytotoxins can accumulate due to the rapid breakdown of the water soluble materials.

4. Straw length – generally microbes will tackle any length of straw but incorporation is easier with shorter cuts. However, there is experimental evidence that chopping shorter than 2 cm may accelerate breakdown.

5. Nitrogen – microbes need nitrogen to break down straw but there is usually enough present in straw and soil for the initial phase of breakdown. Microbes will lock up the nitrogen and release it later as decomposition continues.

6. Other nutrients – microbes will generally find all they need in soil and straw.

The case for adding enzymes, cultures of amended bacteria and fungi or other nutrients to accelerate straw breakdown in soil is unproven under UK conditions. The microbes naturally present in soil should do the job adequately given the right conditions. Microbes play a key role in maintaining soil structure. It has been shown that microbes increase in soil where straw has been incorporated. Therefore in the long term straw incorporation should lead to an improvement in soil structure.

Value of straw residues

It is logical to assume that the incorporation of straw into the soil would improve the organic matter status of the soil which would consequently produce benefits in ease of working and crop yield. Many farmers believe this to be so and there is some field evidence that soils do become easier to work. However, the Ministry view is that, on evidence, there is no direct effect of straw incorporation raising organic matter significantly and there is no apparent effect on yield. In 1984, ADAS reported as follows:

> The products of breakdown of fresh organic matter in the form of crop residues can improve the structural stability of soils. Instability of structure, particularly in soils containing a high amount of fine sand or silt, can occur when soil organic matter falls below 3%. Levels of 2 – 3% organic matter are often recorded in soils under arable cultivations and these soils can exhibit satisfactory structures provided there is a return of sufficient amounts of organic matter from the residues of each crop. The roots of most crops, particularly cereals, can supply this level of organic matter. Results of long term experiments on straw disposal at four Experimental Husbandry Farms in the eighteen years between 1951 and 1968 (when straw yields were less than they are now) proved that this was the case; on four soil types there were no significant differential changes in soil organic matter content on any site. These facts were also reflected in the crop yield responses from this series of trials, since method of straw disposal had no effect. Straw from current average crops of wheat will contain some 15 – 20 kg/ha of phosphate and 25 – 30 kg/ha of potash.

However, work at Rothamsted on a sandy loam at Woburn containing less than 1% organic matter showed that yields of winter wheat, spring barley, potatoes and sugar beet were about 10% greater after adding straw for six years than with basal fertilizers equivalent to the nutrients in straw. The 1984 ADAS conclusion was:

> Burning or incorporation of the whole residues *in situ* will return these nutrients to the soil reserve. Long term experiments on straw disposal at Rothamsted and Norfolk Agricultural Station have indicated that any slight yield increases in subsequent crops in the rotation after straw return could be attributed to these nutrient residues.

Conclusions

Clearly the situation is complex and the final yield results will depend on the very diverse set of circumstances. However, there are clear indications that yields of up to 5% extra can be achieved although we are not quite sure exactly how. Alternatively, yield reduction of up to 10% can also easily follow straw incorporation. The real question now is if it is possible to identify a method of application of the technology which will allow a predictable move to increased rather than decreased yield.

There are three clear areas for study; the management of burned seedbeds, stubbles and incorporation.

BURNED OFF SEEDBEDS

If straw burning is allowed, even within strict limits, it may have distinct, quantifiable advantages as well as problems.

If straw is mixed only into the upper 5 – 7 cm of the soil,

Table 3.4 Effect of methods of straw residue management on the number and size of winter wheat seedlings sown on 29 September after shallow tillage (7 cm) on clay soil

Date of plant count	Straw treatment	No. of plants/m²	Mean plant dry weight (g)
9 Nov.	Burnt	295	0.034
	Stubble only	241	0.026
	Stubble with straw chopped and spread	194	0.023
19 April	Burnt	285	0.84
	Stubble	207	0.70
	Stubble with straw chopped and spread	193	0.63

Source: D.G. Christian, AFRC Letcombe Laboratory.

establishment and early growth of autumn-sown cereals are usually poor (Table 3.4).

The effects are usually more marked in wetter than in drier seasons and are even more pronounced with direct drilling. Possible reasons for such effects include straw affecting the ability of the seed drill to place seed into the soil; the consequences of larger populations of slugs and the effects of toxins from straw decomposing in anaerobic conditions when seed and straw are in close contact. It may well be that tined coulter seed drills, such as the IH 511 or the Moore All-Till will do this job in a way to minimize the problems.

One of the conclusions drawn from Table 3.4 is that the weight of seed required to achieve a given population is less, i.e. seed survival, germination and establishment are better. More seed may well be required after incorporation.

Burning will also reduce the requirements for herbicide and, probably, moluscide for slugs. There is also an effect on disease. Septoria and sharp eyespot tend to be carried on stubble so a good burn will reduce this risk. Normal eyespot and take-all will not be affected by burning. Burning does tend to knock down over-all inoculation levels, but even if burning is carried out, the headlands and firebreaks remain.

If burning or ploughing is planned to reduce the chance of disease transfer, then it should be done before the adjacent field is drilled and has emerged; emergence is the key issue. If ploughing or burning was so long ago as to allow the land to green up, then Gramoxone can be used to destroy the Green Bridge which would carry over disease.

Burning off, where allowed, will have to be followed by ash incorporation. Such a situation may lead to an improvement in yield. Figure 3.2 shows how burning and shallow cultivation lead all other treatments in yield. (Figures average over 7 years from Letcombe.) A similar story is shown by the figures in Table 3.5 where shallow cultivation gave good results where all the straw was burnt.

Table 3.5 Effect of different methods of tillage on yelds of winter cereals and oilseed rape, 1974 – 1983 where straw residues were burnt

	Relative crop yield (ploughing = 100)		
	Direct-drilled	Shallow tillage (5 – 7 cm)	Ploughing
Clay (35% clay)	104	103	100
Clay (55% clay)	101	(not tested)	100
Silt loam	99	103	100

Source: Letcombe Laboratory

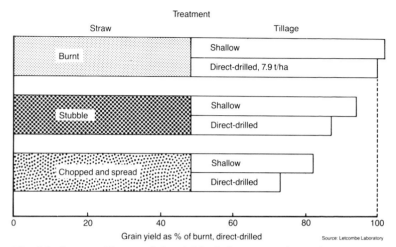

Fig. 3.2 **Crop residues and the yield of winter cereals on clay soils 1976-1982.**

Clearly, under some conditions, burning followed by some form of shallow tillage or a scratch seedbed will give the best results and under many conditions (if not most) give the best weed control.

It is interesting to note the position of direct drilling in these figures. All the ADAS work over the last ten years has tended to show that direct drilling gave the best yields over all other systems more years than not. It is true that the direct drilling of cereals did expand steadily in the 1970s but that growth slowed and reversed early in the 1980s. This was probably due to two factors, grass weeds and an untidiness in the crop during establishment that just did not look right. Grass weeds can, in fact, be controlled using cultivation rotation (see p. 87); there is a good case to use direct drilling for several years, but not indefinitely. The second, untidiness, was, at least in part, due to the dominance of the 3 disc drill. Under some conditions this produced an uneven emergence and establishment. The new generation of tined drills such as the Moore in particular (with a disc and a tine) or the IH 511 drill solve many of the problems. Further development will further improve the drills' ability to produce even establishments in burned off and scratched seedbeds. The scratch seedbed which is necessary with any burned off area will produce the better yields being sought.

Plate 3.6 There has been some interest in 'rotary' coulters to incorporate trash and straw within the top few inches before it is ploughed in.

Procedure on burned off ground

- prepare headlands and fire-breaks (shallow plough)
- burn under control as allowed by law
- incorporate the ash with light surface cultivation
- allow weed seeds to germinate, if significant Green Bridge, burn off with suitable herbicide such as Gramoxone
- follow options:
  direct drill over burn and headlands, using Moore or IH drill (3 disc drills may be acceptable under the right conditions) or cultivate burned area to match headlands, follow minimal cultivation system and drill with conventional drill or deep plough, cultivate and drill.

STUBBLE MANAGEMENT

There is no doubt that the established methods of stubble cultivation
with or without the mouldboard plough do work. Each cultivation
method has its own advantages and limitations. One of the most
compelling reasons for the move to the chisel plough and reduced or
minimal cultivations has been the reduced power requirements, the
increased rate of work and the improved 'system potential', i.e. more
work completed per man per season.

Cultivation – power and output

David Patterson of the NIAE has produced some figures comparing
different systems on the basis of their power requirement (Fig. 3.3),
and their typical output as 'system output' (Fig. 3.4). Clearly, shallow
cultivation (5 to 7 cm) such as discs or light harrows will give high
outputs. Obviously, deeper cultivation (15 to 20 cm) has its price but
some of the recent developments of combinations can achieve

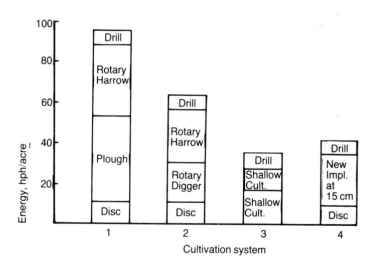

Fig. 3.3 Energy required to incorporate chopped straw (clay soil).

NIAE

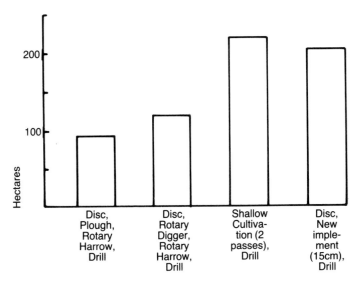

Fig. 3.4 Area potential of cultivation systems (clay soil).

significant improvements in output. One of the questions to be asked is whether this deep cultivation should be done with the mouldboard plough or with chisels. Fortunately, there are some clear indicators as to when to use each type.

The basic logic is that it may help to establish the seedling if surface cultivation is carried out (see Fig. 3.2), i.e. to create a 'seedbed'. It may also be necessary to cultivate deep to allow the continued growth of adventitious roots. There is little argument for anything in between except to undo the bad effects of traffic compaction or to incorporate stubble or other trash.

Nurse stubbles

The real question is whether the stubble needs to be incorporated and what it will cost in timeliness and resources to do so. The indications (see Fig. 3.2) are that it is safer to bury the stubble by whatever means suits the conditions on the individual farm.

Plate 3.7 Traditional methods of stubble cultivation may well be useful for incorporating straw.

The interesting possibility is that the stubble could be used to nurse the next crop in the way that a coarse tilth has been recommended for autumn sown crops. The stubble could protect the seedling and hold the surface open for gas exchange and water percolation. The logic is that there could be an effect in such a way, but the evidence is that any such possibility is limited. Yields from direct drilling into stubbles have generally been slightly depressed. The average depression would be expected to be in the region of 10% (Fig. 3.2). However, if the stubble is given a light cultivation before drilling, that is likely to be reduced to 5%. Two questions now emerge. Firstly, could the yield depression be eliminated or turned into an increase in yield? Secondly, if not, is the value of the yield depression less than the cost of cultivations and loss of timeliness required to do something more drastic? The answer to the first question depends on getting the technology right. The answer to the second depends on the individual circumstances, but the direct cost savings alone (not counting timeliness or system benefits) could easily outweigh the yield reduction.

The best indication at present for good nurse stubble management would be as follows:

- deep cultivate such as Paraplow or Shakaerator if it is known that the soil needs it for good root growth
- light harrow with spiked harrow to create mini-tilth and encourage weed seed germination
- paraquat spray to eliminate the Green Bridge
- drill with tine drill such as Moore All-Till or IH 511

INCORPORATION

The general logic

All the available information (see Effects on yield, p.37) suggests that while there are potentially serious problems, there are clear opportunities to maintain or possibly increase yields under incorporation systems.

There is no doubt that researchers in the UK have found evidence of depressed yields following straw incorporation. We know there are straw toxin problems with germination and establishment. On the other hand, we know that the farmers of Schleswig Holstein do incorporate straw and produce high yielding crops in the 10 tonne club.

Can we produce a set of rules that will give successful straw incorporation under UK conditions, with good yields and more economical power inputs than are accepted by the German farmer? While research is still incomplete, it is already clear we can do it.

Figure 3.5 comes from Brian Sanders of ADAS, who has been to Schleswig Holstein. It is quite clear that German farmers tend to pump in power; roughly twice as much as our reduced cultivations. The system in the histogram is based on mouldboard ploughing – and most farmers throughout Germany plough during cultivations following straw incorporation. It works. Yields are often good and farmers make money.

The UK concern is that there may be lower yields and higher power inputs. Could we get the best of both worlds? Most of the research work done at EHF and research centres run by the Agricultural Food and Research Council have been concerned with the rate of decomposition of straw when turned in as a layer. The main flush of toxins under these conditions will fade within six weeks.

David Patterson, of the NIAE suggests that this period might be as

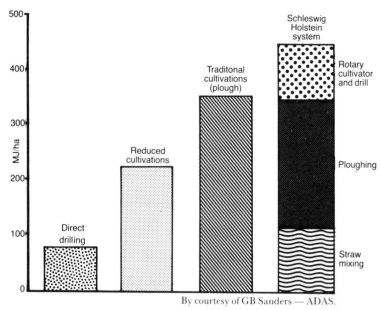

By courtesy of GB Sanders — ADAS.

Fig. 3.5 Comparative energy inputs.

little as two weeks if the straw were chopped really fine and spread out. This view is confirmed by Dr Karl Koeller at the Rhineland Chamber of Agriculture. Dr Koeller states that it is possible to incorporate up to 10 t/ha of straw without loss of yield, provided that there is thorough mixing.

In fact many German farms plough in a layer of straw and soil moved with trash boards from the surface between two furrow slices. The best conclusion from the research and field experience appears to be that this must be limiting on the following crop. The conclusion reached from putting together Dr Koeller's conclusions and research from the UK, France and the University of Copenhagen, Denmark, appears to be that we need to spread the straw as evenly as possible throughout at least 15 cm (6 in) of soil.

Assuming that conclusion to be sound, the practical question is how to go about doing it in the field. In theory, the rotary cultivator is a very good mixer. Mr Patterson is convinced of the practicality of a rotary digger provided that it is man enough for the job. Dr Koeller says that German farmers do use these tools but work rates are slow.

Long term German field trials have tended to favour discs and rigid tines. Discs are used in Germany and Denmark on light soils where the straw crop is not too heavy. German farmers on the whole prefer to use fixed tine, heavy duty cultivators because they give good incorporation to the required depth, high work rates and low running costs.

It will be necessary to have at least two passes with the cultivator and then a delay of at least two weeks, preferably longer, before drilling. At drilling, a powered cultivator can be used to deliver a seedbed in front of the drill; here the German farmer has tended to move to the combination of harrow/drill.

The German farmers tend to use Suffolk type coulsters on ground that has been mouldboard ploughed. In Schleswig Holstein, it appears to be common practice to incorporate into the surface, plough and drill into a trash-free seedbed.

Reducing power inputs for UK conditions looks possible but we may have to look again at drill designs to see how these can work in seedbeds incorporating some partly decomposed straw.

Work at the NIAE has shown that the Moore drill appears to be able to cope well; we know that certain types of coulter fitted to the Amazone drills have a track record on combinations used by German farmers. It may well be that some of the other drills we already have in the UK will show up well. The fashionable air seeder drills should cope well with straw in their seedbed but most look as if their uniformity of depth of placement must be suspect. It is clear that we still have something to learn about coulter design.

On the subject of combinations, these do have something to offer but it may well be that having a trailed drill behind the harrow may successfully reduce the power requirement which is needed just to lift the drill.

Moisture retention is very much in the minds of farmers in Germany, France and Scandinavia. This is one of the reasons for going to combinations of tools working behind the same tractor. Where implements are on different tractors, they tend to operate in the same field, one behind the other, with outputs matched.

Compacting the seedbed with crumble rollers is also common practice. Press wheels on the plough are commonly seen. Good contact between soil and straw is all part of maximizing the speed of breakdown and helping a more even establishment.

There have been a number of ideas that have cropped up in the last couple of years. One was Harvest Services' secondary knife for mounting behind the combine table. This would allow minimum straw through the combine and minimum stubble length. Another was some really heavy discs such as the Super Utah from Parmiter, with working depths from 10 to 20 cm. Bomford also have a straw screen on their Dynadrive.

Incidentally, British cultivator design is probably the best in the world. If you go for winged feet, look for at least 4 in of lift to get any reasonable effect. Haylocks, for example, would work well in straw incorporation. The McConnel, Shakaerator/Tillaerator combination may also perform well with incorporation.

There is one other relevant point. The option to change the system year by year is an important one and straw incorporation must be fitted into that concept. Some drills may well be able to direct-drill into short cut stubbles. We may need to minimal cultivate with straw and, in other years, we may have other reasons to mouldboard plough. There is not going to be a rigid set of rules for straw incorporation.

The real key to the whole operation appears to be rapid and thorough mixing of finely chopped straw. Thorough mixing into 15 cm (6 in) of soil may well allow drilling without ploughing fairly soon

Plate 3.8 Volunteer cereals appear to grow quite well immediately after the incorporation of quite large volumes of straw. If this is so, why should we not be able to plant crops into this situation?

afterwards. Moisture retention and compaction of the seedbed look to be important. Drill design is something we may be on the right track with, but we still have something to learn.

Overall, there is clear evidence that we can do this job satisfactorily. We need more experience on our soil types; but field experience is beginning to show the way we have to go.

CHOPPING

Chop length
In a nutshell, the problem with chopping on the combine is the stubble. Maximum decomposition rate for incorporated straw comes from a short chop. Stubble is often 10 – 15 cm long and tends to be sticking out of the soil anyway. Conclusion? Chop the stubble. If we are going to do that, why not spread the straw off the combine and chop the whole lot on the ground? That would also free the combine from the extra power demand and complexity.

The alternative argument is that chopping the straw on the combine involves one less pass, is much more convenient, and involves less power. German research shows just how much power is involved in combine chopping (Table 3.6).

Table 3.6   Combine mounted choppers

	Dry throughput t/h wheat straw	*Hp required*
	4	11.6
	8	18.7
	12	25.2

Source: DLG test 2243

At 1984 prices, the combine chopper will cost £1.30 – £1.50 per acre to run. A typical contract charge to run a separate chopper behind a tractor a would be £10 – £11 per acre.

Work at Letcombe has shown the following figures for the decomposition rate for straw on one particular experiment.

Milled to less than

1 mm took 14 days
0.5 cm took 29 days
1.0 cm took 30 days } to lose 50% of wt
2.0 cm took 47 days
5.0 cm took 54 days

We need short chop length of all the straw. That includes the stubble. We also need even spread and that is not easy to achieve with a combine chopper. Suppose we spread the long straw and then follow behind with a chopper that could cover 3 or 4 m in one run and chop everything, including the stubble, to a short length. Incorporation with tines and discs would then be easy and decomposition would be rapid. The important part of the decomposition is, fortunately, the early part when significant levels of toxins are produced. The key, then, is to chop the straw, all of it, to as short a length as prudent use of energy will allow. Immediate incorporation with the straw firmed up to get good soil/straw contact is also important so a crumble roller on the back of the incorporator appears to be a good idea.

The Ministry view assumes that chopping is certainly desirable to make incorporation easier from a mechanical point of view, but points out that the requirement for rapid breakdown is only a logical need – there is no direct link with yield.

Straw Utilisation and Disposal, MAFF 1984 stated the position as:

> Short straw lengths, well lacerated can be incorporated and decomposed more readily than longer intact straw. The chopping operation can either be done from the swath or by direct mounting on the combine. Current mounted choppers will normally produce a range of straw length in the chopped straw, but seldom are the maximum lengths less than 12 – 15 cm.
>
> To achieve the ideal of less than 5 cm would require much more power and would be likely to slow the combine operation. Chopping from the swath is not generally favoured and purpose built machines currently available do not achieve a short chop. The use of double chop forage harvesters to produce less than 5 cm lengths is being considered by some farmers. A secondary cutter bar on combines to shave stubbles is another recent introduction to aid incorporation. Axial flow combines appear to break up straw more than conventional machines.
>
> Barley straw is more difficult to chop than wheat because it is less lignified and so has a lower stiffness. Similar differences in chopping performance have been observed in the field between different wheat varieties but these have not been scientifically quantified. Maturity will affect the brittleness as will incidence of disease. Avalon and Hobbit appear to break up readily and therefore are easier to incorporate and decompose. The degree of pre-harvest weathering will also be a factor.

In the same year of 1984, ADAS and the NIAE started a large survey of straw choppers, due for publication in Spring 1985. The results of that survey compared performance of machines in the 1984 harvest. Design changes and developments were, of course, taking place rapidly at that time. New generation choppers of both the on-combine and tractor operated types will continue to alter the position.

System and depth

All the discussions so far indicate that several systems could work. Deep cultivation is only necessary for normal plant root growth where a pan or compaction needs to be eliminated.

Figure 3.6 shows results of work in Germany and collected by the NIAE. It shows that depth does help to rot straw, but the graph levels off at about 20 cm depth. This would be on a soil with relatively low clay content compared with UK conditions. There is little evidence that going deeper than this would help at all and some evidence that at greater depth rotting would actually slow down. This is probably marked on clay soils. Incorporation, then, more than 15 cm (6 in)

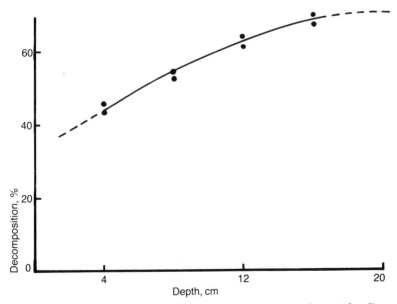

Fig. 3.6 Effect of depth of straw incorporation on straw decay (after 7 months) – *Hohenheim* 5.0 t/ha chopped straw.

deep is not necessary. Ploughing, therefore, at 15 – 20 cm (6 – 8 in) seems quite adequate. Chisel cultivation to the same depth also appears to satisfy the demands of the known technology. It is relevant to ask if this depth could be reduced and there are some indications that it could.

Table 3.7 shows the 1984 results from Bridgets EHF:

Table 3.7 Preliminary results immediately after harvest 1984 (Bridgets EHF – Winter Wheat)
(Grain yield in tonnes/ha corrected to 15% m.c. Straw incorporated at 4.5 tonnes of baleable straw/ha estimated at total straw yield of 9.0 tonne/ha)

| | Plants/m² | | | Fertile | |
	Autumn 1983	Spring 1984	Ear nos/m² Jly '84	ears/plant Jly '84	Yield
Plough 20 cm	280	190	604	3.1	8.74
Plough 10 cm	285	194	535	2.7	8.75
Tine 15 cm	247	184	603	3.2	8.44
Tine 10 cm	237	171	600	3.5	8.48
Rotadigger	266	174	662	3.8	9.09
Dynadrive	204	142	556	3.9	8.31
Tillaerator	221	149	630	4.4	8.52
Rotavate and plough 20 cm	270	186	526	2.8	8.77
Burn and direct-drill	263	280	625	2.2	9.19
Burn and plough 20 cm	298	193	510	2.6	8.84

Least significant is 0.72 tonnes per hectare, i.e. *none* of the yield differences are statistically significant.

Firstly, column 3 shows that there was a fair level of casualties of plants over-winter. Secondly, the winter wheat did show a remarkable degree of compensation. Incidentally low populations in the autumn of 1983 were largely associated with straw in the surface of the seedbed.

By the time of harvest 1984, the differences between plots had evened out and the differences in final yield were relatively low. Bridgets' Ben Freer points out that the differences were not statistically significant but they do show trends which were confirmed in 1984, having had similar results in 1983. Clearly, there are many factors involved but at least on the Bridgets' chalk soils it was clearly possible to achieve just as good a yield by incorporating relatively shallow. The straw yield was estimated at 9 tonne/ha and, of course, on a chalk subsoil, drainage and subsoil would be in good order. The conclusion, then, would appear to be that if the subsoil is in good order, it is possible to work shallow and quickly. This keeps work rates up and costs down.

If you do opt for the mouldboard plough, then the figures show that shallow work can be entirely satisfactory. The general consensus is that to incorporate with the plough, it is necessary to use trashboards. Most manufacturers take the view that you have to plough deep enough for the body to go in deep and bring the trashboard into action. This problem can be got over in two ways. Firstly, both Colchester Tillage (with their Lemken range) and Krone have purpose built shallow ploughs with trashboards as a standard option. These will plough 10 – 15 cm (4 – 6 in) deep. Alternatively, McConnel's who import the Fiskars plough from Finland say their deeper bodies have adjustable trashboards that can be set so that

Plate 3.9 The McConnel Discaerator uses discs and deep chisels to incorporate straw quite well.

whatever depth the plough is worked at, the trashboards can be brought into effective action.

Leaving the straw near the surface allows aerobic destruction of the straw which is faster and has reduced toxin problems.

One trap that we must not fall into with the mouldboard plough is to think that it solves the weed problem. It may help in the short run, but ploughing to the same depth next year may well perpetuate the problem. Light cultivations may well be a cheap way of killing rather than burying weeds. That will certainly work in the dry but in the wet, cultivation may merely transplant weeds. In such a situation, covering the ground relatively cheaply and effectively with a blanket weedkiller such as Gramoxone is a technique which must be part of the overall strategy.

Soil type and energy inputs

In all the ADAS work for ten years or so, on the heavy soils where most of our cereals have been grown, direct drilling has come out best on more trials than not. Table 3.8 shows the results of 126 ADAS field experiments between 1972 and 1982. On these trials, the technique was, of course, used according to the proper rules.

Table 3.8 Grain yield – various systems compared

	Plough deep cultivation	*Shallow cultivation*	*Direct-drill*
Winter wheat			
heavy land	19	21	32
Winter barley			
light land	10	3	1
heavy land	2	3	2
Spring barley			
light land	7	5	1

Number of sites with highest yield
Clearly on heavy soils, direct-drilling had definite attractions even if it also had its limitations. On light soils, there was no such yield advantage and deep cultivation and ploughing was always much more attractive.

A further disadvantage of moving back to the plough on heavy soils is the cost of ploughing. It is obviously more difficult to plough than a sandy soil (Table 3.9).

Table 3.9 Plough-draft (Ref. Hunt of Illinois)

Silty clay	13 lb/in^2
Loam	6 lb/in^2
Sand	4 lb/in^2

Table 3.10 Energy inputs

	Energy input index
Direct-drill after burn	1
Shallow cultivation burn	2½
Plough and incorporate	6
German system – incorporate and plough	9

Source: ADAS 1983

The energy input of straw incorporation systems based on the mouldboard plough is not attractive (Table 3.10).

Clearly then, the attractiveness of the mouldboard plough on light soils is relatively uncomplicated. On heavy soils, the disadvantages become much more compelling. One answer which applies to all soils is to plough relatively shallow. Trials run by ADAS in 1984 for 1985 harvest were laid down to look at the effectiveness of different depths of ploughing from 12 to 23 cm (5 to 9 in). The indications are that working 15 – 20 cm (6 – 8 in) deep will be quite adequate.

One alternative is to go to discs and chisels which may have higher rates of work and be easier to use on the heavy clays where several passes may be necessary. A further possibility is to utilize the straw via the baler and go to nurse stubbles.

Nitrogen

There has been discussion for some years about whether extra nitrogen fertilizer is necessary or helpful in accelerating the breakdown of incorporated straw. One thing is clear, in soils which are overall well manured, the effect of extra nitrogen is small.

One of the problems in the research is that 10 – 15 years ago less fertilizer (including nitrogen) was used overall. A response, then, to extra nitrogen to help the straw rot down was more likely. Further, it was often said that extra nitrogen was necessary only in the first year or two. In practice straw will lock up nitrogen and release it slowly as the material breaks down.

Figure 3.7 shows the results of research by Bent Christensen of the

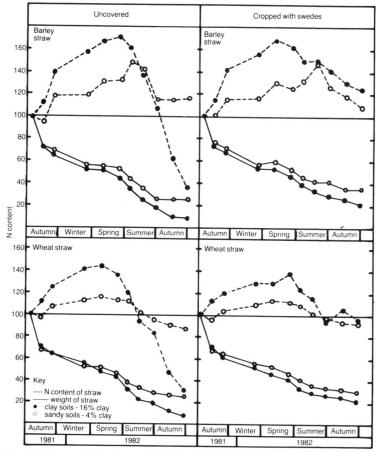

Fig. 3.7 Immobilization of nitrogen.

Danish State Agricultural Station, Ronhave. The graphs show the immobilization of nitrogen. The dotted lines show the nitrogen content of the straw on clay (16% clay) and sandy (4% clay) soils. The soil nitrogen is immobilized over the winter period and then slowly released as soil temperatures rise. The release of nitrogen in the spring is slowed down if the land is cropped. An interesting aside to this observation is that fallow land clearly allows more loss of nitrogen by leaching.

Studying the shape of the graphs makes it clear that for autumn crops there is a reduction in the amount of available nitrogen and the implication is that extra may be useful for autumn grown crops. It is

Site	Clay %	Straw						
		Removed		*Incorpor'd*		*Burned*		
		Normal N	*Extra N*	*Normal N*	*Extra N*	*Normal N*	*Extra N*	
(c)	4-7 33,3	+0,9		+1,2	+0,7			1974-82
(a)	7 34,9	+0,3		+1,6	+1,0	+0,5	+1,0	1967-82
(a)	12 44,2	+0,9		−1,0	−0,2	+0,1	+0,2	1967-82
(c)	11-13 48,7	0		−0,8	+0,4			1974-82
(a)	16 51,0	0		−0,1	+0,8	+0,6	+0,7	1967-82
(b)	25 50,8	−0,3		+0,3	+0,1	0	+0,2	1968-78

Normal nitrogen for Spring Barley
Extra nitrogen provied 40 kg extra N iin the autumn

Fig. 3.8 Effect of extra N on yield after incorporation.

also clear that sandy soils do not release their available nitrogen as easily as do the clays. Clays may need extra in the autumn but let it go more easily in the spring. This mechanism may give at least marginal help with reducing nitrogen run off by helping to lock it in over the winter period.

Figure 3.8 is also from the Danish research station at Ronhave. Generally, extra nitrogen in the autumn did result in extra yield the following year but the increases were small. The Danish conclusion, even on their relatively light, infertile soils, was that it was difficult to justify extra nitrogen on economic grounds.

The current ADAS view is that on present evidence there is enough mineral nitrogen in soils after harvest to ensure that nitrogen is not limiting either rate of straw breakdown or yield of the next crop. It is possible that on some sands where soil nitrogen supply is lower, this is not true and that some autumn nitrogen would be justified. However, even then there is no case for exceeding 30 kg/ha. Generally, the ADAS view is that autumn nitrogen is economically and environmentally not justified.

Trials at Boxworth were more definitive, moving the argument away from extra nitrogen. The figures in Table 3.11 (a) tend to show marginal increases in yield for extra nitrogen but they are all very small and not significant in economic terms.

Straw was chopped with a combine mounted chopper on 13 August and was either burnt or incorporated into the soil from 24 to 31 August. The secondary cultivation with medium discs was done on 18 September. Paraquat was sprayed on 22 September and 50 kg/ha N applied to split plots. Slug pellets were spread in early October. A second paraquat spray was applied before drilling.

Table 3.11 (a) Boxworth EHF: Trial of incorporation and extra nitrogen. Results 1983 (first year); yield (t/ha) at 85% DM; effect on yield of straw incorporation methods

| Treatments | *Autumn nitrogen (kg/ha)* | | |
	Nil	*50*	*Mean*
	(SEDs T 1 – 6 ± 0.167)		(0.130)
	(SEDs T 7 and 8 ± 0.118)		(0.113)
	(SEDs T 1 – 6 vs 7 and 8 ± 0.144)		(0.092)
1. Straw burnt			
direct drill	9.64	9.64	9.64 ⎫ 9.67
5 cm tines	9.55	9.84	9.70 ⎭
2. Plough 20 cm	9.59	9.55	9.57 ⎫ 9.53
+ preplough discs	9.44	9.56	9.50 ⎭
3. Tines 10 cm	9.49	9.58	9.54 ⎫ 9.51
Tines + discs	9.54	9.40	9.47 ⎭
4. Tines 15 cm	9.45	9.38	9.42 ⎫ 9.49
Tines + discs	9.50	9.60	9.55 ⎭
5. Tines 20 cm	9.58	9.40	9.49 ⎫ 9.45
Tines + discs	9.43	9.38	9.40 ⎭
6. Glencoe Soil Saver			
× 1 pass	9.57	9.46	9.51 ⎫ 9.57
× 2 passes	9.53	9.75	9.64 ⎭
7. Rotadigger	9.52	9.49	9.50
8. Mulch Train	9.33	9.44	9.39
		(SED ± 0.037)	
Mean	9.50	9.52	

SE per main plot (cul method) at 32 df = ± 0.159 t/ha or 1.7% of GM.
SE per sub plot (nitrogen) at 33 df = ± 0.180 t/ha or 1.9% of GM.
(T = treatments)
The mean yield after burning was 9.67 t/ha and the mean reduction from all incorporation treatments was 0.17 t/ha, ranging from 0.10 to 0.28 t/ha. Nitrogen applied in the autumn to counteract possible nitrogen immobilization had no worth while effect. Soil nitrogen residues appeared adequate and differences in the soil were small. Amounts of nitrogen in wheat plants were analysed in early spring and amounts were unaffected by the extra autumn nitrogen.

The 1984 results at Boxworth did, however, suggest a marginally different conclusion. Table 3.11 (b) shows the 1984 harvest results. In this case, 40 kg/ha of extra nitrogen in the autumn yielded an extra 0.50 tonnes/ha which was significant. In fact, the response was at least partly due to the fact that, in that year, the total of 200 kg/ha applied in the spring was considerably below the optimum in a year of high yields and good nitrogen response.

The overall conclusion, then, still casts some doubt on the economic advantage, in most years, of extra autumn N to help break down straw residues.

Table 3.11 (b) boxworth EHF: Trial of incorporation and extra nitrogen. Results 1984 (second year); yield t/ha at 85% DM

Treatment	*Autumn nitrogen:*	*Nil + 40 kg/ha*	*Mean*
	(SEDs VI= ±0.205 H only= ±0.149)		(SED ±0.176)
Straw burnt			
1 (a) direct drill ⎱ 　 (b) 5 cm tines ⎰	10.80	11.26	11.03
Straw chopped			
2 Tines and discs 10 cm	9.83	10.18	10.00
3 Tines and discs 15 cm	9.97	10.31	10.14
4 Plough	10.10	10.84	10.47
5 Tines and discs 5 cm	9.66	10.09	9.87
6 Tines and discs 20 cm	9.93	10.27	10.10
7 Pre-incorporation + plough	10.10	10.76	10.43
8 Glencoe Soil Saver	10.16	10.51	10.33
9 Rotadigger	10.02	10.67	10.35
10 Tillage Train	10.16	10.77	10.47
	(SED ±0.047		
Mean	10.07	10.57	

SE per main plot (cultivation method) at 18 df = ±0.215 t/ha or 2.1% of GM.
SE per sub plot (nitrogen) at 40 df = ±0.257 or 2.5% of GM.

Spring nitrogen was split with 45 kg/ha on 9 March and 160 kg/ha N on 30 April. An MBC eyespot fungicide carbendazim at 0.25 kg a.i. (Derosal) and chlormequat at 1.75 l/ha were applied on 28 April, followed by triadimefon + captafol (Bayleton CF) at 2.5 kg/ha on 13 June at GS58. An aphicide, Aphox, was applied on 28 June. Harvesting was on 13 August.

Straw amounted to about 7 t/ha at harvest; about one quarter of this was left as stubble about 15 cm long, the remainder was chopped. Three quarters of the chopped straw was 10 cm or less in length, and with the stubble was either burnt or incorporated into the soil.

Incorporation methods left a range of surface trash: ploughing most effectively buried all straw, the Rotadigger most effectively mixed the straw into the soil but left varying amounts of straw (12 – 27%) much of which was the stubble component, on or near the soil surface. The timely primary incorporation of straw at the end of August before wet weather set in enabled cultivations to be concluded without blockages.

Acetic acid formed during the early stages of straw decomposition had declined or was dispersed before the delayed sowing on 20 October. Weather was excessively wet and toxicity effects from decomposing straw appeared to be minimal. Surface traps during the wet weather at the end of September and early October indicated

Plate 3.10 Wide sets of discs can be very useful with surface incorporation.

high slug activity particularly on plots with little or no surface trash following the use of the plough or Rotadigger. However, after overall treatment with slug pellets there was little evidence of damage.

There is some logic in terms of where to add extra nitrogen. Applying the fertilizer in a form where it would be close to the straw should have advantages. Application as a liquid sprayed on to the straw before incorporation is an obvious possibility. There is, however, not yet any conclusive evidence that such an approach does have any effect on the rate of breakdown, the production and release of toxins or the growth and yield of the subsequent crop. However, these are limited pointers.

Work by INRA (Institut National de la Recherche Agronomique) in France, has shown that when vinasse, a residue from industrial alcohol manufacture, was applied to wheat straw before incorporation, yield of the following sugar beet crop was slightly increased. Other materials such as slurries of animal waste containing readily available nutrients can stimulate decay, especially in dry conditions. Some farmers with large amounts of slurry to dispose of are successfully incorporating their straw after applying the slurry.

There is one other possibility and that is that the nitrogen content of the straw itself can be altered by treatment of the crop while it is growing. It is known that adding late nitrogen to a crop alters the nitrogen content of the grain. There is some evidence that this applies to the straw also. If this were so it is possible that this would speed up the rate of decomposition of the straw.

Results from Broom's Barn and Jealott's Hill have shown that overall application of N to the growing crop does affect N content of the straw as well as the grain. It is now established, of course, that applying late N in, say April, can help lift the N content of the grain and help put it into a quality premium bracket. It is logical that, if that happens, so should the N content of the straw rise. This is, in fact, the case as shown by the figures.

Relation of Nitrogen content of straw to fertilizer application

Broom's Barn			
	90 kg N	180 kg N	330 kg N
N% in straw	0.35	0.43	0.75
Jealott's Hill			
	50 kg	125 kg	200 kg
N% in straw	0.56	0.76	0.85

These figures show that raising N application overall does raise the N content of the straw. There is a question mark remaining as to *when* to apply extra N. In another trial at Broom's Barn, a wheat crop was given 250 kg of N in total in the 1984 season. The dressing was split 50 kg early March, 100 kg late March and 100 kg in late April. A further 50 kg was applied in the late April dressing to a further treatment. So one treatment was to apply extra N and apply it late. This treatment:

- increased yield from 10.02 to 10.31 tonnes/ha
- increased N in grain from 2.23% to 2.71%
- increased N in straw from 0.39% to 0.42%

The yield increase was economic and the increase in grain N significant. So on both counts of yield and grain quality the treatment was worth it. The increase in straw N was, however, less significant.

It is clear, however, that N content of straw can be influenced by fertilizer policy. It is logical, but not proven, that increased N content in straw would improve rates of breakdown in the soil and possibly change toxin production. The next step in the logic is that this might affect the yield of the next crop or at least the costs and possibly timeliness of the cultivations necessary to incorporate between crops. This is an interesting area for discussion and further investigation.

WEED CONTROL

There is good evidence that many farmers will be able to use the mouldboard plough to incorporate straw. Using a shallow plough with trashboards may be a flash of vision on some soils. We do not need to plough deep but we do need trashboards in action to mix the straw.

There is a further advantage in ploughing that some herbicides may work better on plough as Table 3.12 shows. The chemicals tested did show better percentage kills of blackgrass after the plough than after chisels (which was better than direct-drilled plots), further herbicide may work better on plough (see Table 3.12). This effect is less pronounced if the straw is burnt. Suppose, however, we want to use chisel cultivations to incorporate straw and kill weeds. Could this do the job and give a satisfactory crop? The indications are that fine chopping of all the straw including the stubble will allow rapid breakdown as soon as the straw is in the soil with some moisture. So, rapid incorporation immediately after the combine is desirable. The right combination of chisels and discs will get a good mix. What effect will this have on the weed population? There is some evidence from the Weed Research Organisation that seeds in a seedbed with straw in it will be inhibited and take longer to germinate. Look at Table 3.13. It is clear that blackgrass tends to emerge *before* drilling after

Table 3.12 Effects of cultivations, method of straw disposal and herbicides on the percentage mortality of blackgrass

Timing	Pre-emergence				Post-emergence	
Herbicide	Chlortoluron		Isoproturon		Isoproturon	
Rate kg/ha	1.60	3.20	1.14	2.27	1.14	2.27
Straw baled						
Plough	59	95	55	73	99	100
Tine	56	67	10	45	99	100
Direct-drill	16	54	15	33	97	99
Straw burnt						
Plough	64	95	64	57	95	100
Tine	69	94	59	80	98	100
Direct-drill	11	74	−11	44	98	98

Source: WRO

burning but *afterwards* if straw is left about. If we are not going to burn, then Table 3.12 gives the argument to plough. Coming back, however, to higher rates of work and non-ploughing systems, what are the implications? Clearly, from Table 3.12 there is an obvious conclusion that we will rely heavily on post-emergence herbicides. However, if such post-emergence herbicides are soil acting there will still be a problem on seedbeds containing high ash or organic matter. This has its problems. Look for a moment at pre-emergence herbicides. Table 3.12 shows that with pre-emergence herbicides, the presence of straw reduces the effect of soil applied compounds. Again from WRO work we know that the ash build up from successive years of burning will absorb and inactivate soil applied chemicals. So, the pressure seems to be against pre-emergence sprays and in favour of post-emergence application.

Table 3.13 Effect of straw disposal system on the proportion of A. Mysuroides plants recorded before and after direct drilling: Autumn 1977

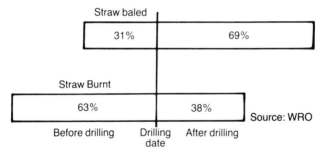

There are, however, problems. Soil and crop conditions may not favour spraying in the crop. Spray application rates will have to be high to achieve a good kill, which is expensive and may produce problems with phytotoxic damage to the crop. In any case, the philosophy of control is to attack early to reduce competition.

So we have the approach of minimal cultivations. Certainly it will get the energy input down and the rate of work up. Table 3.14, however, shows what happens if we cultivate germinated blackgrass and desiccation does not kill the seedlings. In effect, we transplant them and the number of flowering heads in the crop at harvest is multiplied up.

Table 3.15 shows what happens as cultivation is delayed. The numbers of plants of blackgrass tends to go up because of transplants

Table 3.14 Blackgrass transplants growth stages

	Dec. 1978	April 1979	July 1979
Blackgrass emerging after drilling	1 leaf	1 tiller	5 heads/ plant
Transplants	3 tillers	6 tillers	22 heads/ plant

Source: WRO

Table 3.15 Effectiveness of spring tine cultivations at destroying blackgrass prior to drilling

	Dates of cult. and drilling		
	11 Sept.	2 Oct.	24 Oct.
BG present before cult. (plants/m^2)	97	138	151
Transplants in crop (plants/m^2)	6	29	47
% control by cultivation	94%	79%	69%
Growth stage at cult.	2 leaves	2 tillers	5 tillers

Source: WRO 1979

if the kill conditions are not ideal. So, the key is an early attack or a totally complete kill probably with chemical just before drilling. Here is the flash of vision. Ten to fifteen years ago, the WRO was advocating the use of chemicals such as Gramoxone and cultivations to progressively weaken a weed population every time the ground 'greened-up' during the post harvest/pre-drilling period.

Table 3.16 shows work from Steve Moss at WRO showing the use of chemical and cultivation in the control of blackgrass. The system, then, is straightforward. Get in behind the combine and break up the ground with chisels and discs slung on the back. Look at the life cycle of the weeds involved and as soon as there is a little green out of the seed, one full leaf, hit them. If it is dry, cultivate with spring tines or anything cheap. If it is slightly wet, spray cheaply with paraquat (Gramoxone). In practice, cultivation does not work well because it is not dry for long enough. Chemical sprays are a much better insurance against weed competition in the following crop.

For those farmers who do not want to mouldboard plough, those who want to chop and chisel, we have the basis of a system of weed control. It may be necessary to rotational plough say one year in five depending on the situation. In most years, higher rates of work, lower power inputs and better timeliness can probably be achieved without loss of yield, by the approach described using light cultivations and sprays in a combined attack before drilling with careful use of post emergence spraying.

Table 3.16 Blackgrass transplants

	April 1979		July 1979
	BG emerging Post-drilling (plants/m²)	Transplants in crop (plants/m²)	(heads/m²)
Paraquat + cults	98	0	364
Cults only	117	50	882

Source: WRO

DISEASE CONTROL

There is no doubt that the safest thing to do is to burn. The presence of straw in the seedbed or on the surface does increase the amount of inoculum present.

In the case of direct drilling into 'Nurse' stubbles, straw stubble may increase the likelihood of certain diseases. The incorporation of a full crop of straw may add significantly to the inoculum present especially if straw is left in the surface. There is no doubt that the cultivation system does affect disease levels and that the presence of trash does encourage trash-borne diseases.

Experiments on different cultivation treatments were carried out at Rothamsted in 1974 and 1975 in collaboration with the NIAE. The incidence of three soil-borne diseases and the yields from the two winter wheat sites are given in Table 3.17 as means for the two years 1974 and 1975. Cultivation treatments generally had only a small effect on disease incidence; both take-all and eyespot tended to be less after chisel ploughing or direct drilling than after ploughing, but at Boxworth in 1975 the difference was much larger and only 37% straws were infected by eyespot after direct drilling compared with 74% after ploughing. These decreases in disease were not reflected in the yields which were largest after ploughing. In 1974 no foliar diseases were prevalent at either site, but in 1975 yellow rust was severe on Maris Templar at Boxworth (24% area of second youngest leaf infected); there were no differences between cultivation treatments (R. Prew, Rothamsted).

ADAS work has given considerable insight into how plant diseases will be affected by incorporation.

Table 3.17 Effects of cultivations on soil-borne diseases and yields of winter wheat (means of different secondary cultivations and of two years 1974, 1975)

Primary cultivation treatment	Boxworth EHF				Rothamsted			
	% Straws with eyespot	% Straws with Fusarium foot rot	% Plants with take-all	Yield (t/ha)	% Straws with eyespot	% Straws with Fusarium foot rot	% Plants with take-all	Yield (t/ha)
Plough	70	31	23	6.24	70	17	58	5.66
Shallow plough	78	34	22	5.95	68	20	60	5.34
Chisel plough	68	35	21	6.04	64	22	51	5.31
NIAE rotary digger	67	42	30	5.81	72	29	64	5.32
Rotary digger	71	37	24	6.05	65	23	63	5.42
Direct drill	50	35	16	6.05	64	24	54	5.33

The Green Bridge

Some pathogens require a live host for carry-over. The most important pathogens on cereals in Britain are barley yellow dwarf virus (BYDV), yellow rust, the brown rusts and powdery mildew. Yarham and Hurst (ADAS 1973) pointed out that direct drilling relies greatly on the herbicide being completely effective but this may not be achieved when rain falls soon after spraying, or when seeds survive spraying in dry weather but germinate later. The surviving 'volunteer' plants may provide foci of diseased plants bridging the gap between sown crops and so establishing early and severe attacks in autumn drilled cereals. This underlines one of the reasons why scratch tillage before direct drilling often improves results; seeds are encouraged to germinate and can be killed before the next crop is planted and established.

Plate 3.11 Scalloped discs are helpful in dealing with straw.

There is another problem created by early drilling which can, again, create a green bridge. 'Reduced cultivations and especially direct drilling are quick compared to ploughing so that brief dry spells can be utilized by the farmer. This encourages early drilling and the early sown crops emerge when vectors and spores are still plentiful and temperatures are high enough to encourage infections and incubation. Thus early sown winter oats have more BYDV and yield less than crops sown later' (observed by Plum in 1974).

In July 1984 ADAS published the following summary of the situation.

Trash-borne diseases

Non-ploughing techniques might be expected to aggravate eyespot infections because they leave more straw bases on the soil surface, but this has not always proved to be the case. Results from Boxworth EHF and Letcombe Laboratory (1973 – 1975) indicate that although reduced cultivations may initially increase the incidence of eyespot infection, direct drilling can also reduce eyespot. This phenomenon is thought to be due to an increase in (antagonistic) microbial activity in surface layers of the soil. This is affected also by previous cropping. A further complication is that deep sown plants with an erect habit appear to be more susceptible to eyespot than shallow drilled prostrate plants. Comparison of baling, burning and chopping of straw on ploughed, minimum cultivated and direct drilled plots at Boxworth and Drayton EHFs suggested that there will be little to be gained in terms of eyespot control by straw burning. Stem bases infected with eyespot produce large numbers of spores; one straw can infect an area of 10 m^2 of crop, so burning or burial by ploughing has to be extremely efficient to have a marked effect on eyespot incidence.

Rhynchosporium and septoria are trash-borne diseases which one would expect to be favoured by non-ploughing or non-burning. There is some indication of this happening with rhynchosporim at Rothamsted Experimental Station (1974 – 1976) and for septoria at a number of sites (1973 – 1974). In some cases the differences between ploughing and non-ploughing treatments are apparent early in the life of the crop but have disappeared later in the season. This can often be explained either by cross infestation between plots or by weather conditions being of over-riding importance in disease development and thus obscuring initial differences. Sporulation may be increased by the use of herbicides such as paraquat.

Net blotch is a disease which has risen from obscurity to

Plate 3.12 Spading machines are slow but do quite a good job on incorporation.

prominence in the last few years. Amongst a number of recent changes in cropping practice, which are thought to have contributed to the build up of net blotch, are direct-drilling and early sowing. Straw and leaf debris, particularly from the upper part of the plant, is an important source of infection. There is circumstantial field evidence pointing to surface straw as a major source of net blotch infection and it is thought that only one infected straw is needed to infect all the plants in 1 m². Thus straw disposal may be more important for net blotch than for rhynchosporium which develops more slowly than net blotch during the winter.

Take-all can be very serious on light calcareous soils but can also affect crops on heavy soils. Overall, no consistent effect on cultivation method has been observed but the results from individual sites have been very variable. Soil physical conditions and the place in the rotation have an over-riding effect on the incidence of take-all. The high level of debris in the top few centimetres of direct drilled soil was not reflected in a high level of infectivity in this layer, suggesting that there was microbial activity antagonistic to take-all.

Conclusions

Clearly, the burn has attractions from the disease point of view. Stubbles or chopped straw will provide an important source of inoculum of trash-borne diseases.

Where trash is present, from a disease point of view, there is some logic in using the mouldboard plough to bury trash to leave a clean seedbed. Burning off the green bridge with Gramoxone before ploughing will be a further aid to the control of aphid and diseases such as BYDV and several rusts.

A question still remains over the desirability of incorporation before ploughing. There is some evidence that such a practice is likely to spread take-all. If there were not pre-incorporation, it would need good work with skims and trashboards to get the plough to do some mixing. Mixing can logically be expected to help breakdown although there is little hard evidence to support this theory. If take-all is known to be a problem, the use of surface cultivation, however desirable, from a weed control or incorporation point of view, may be in conflict with disease control.

With non-inversion techniques, incorporation of chopped straw may lead to disease problems unless thorough working is used to ensure rapid progress through at least the initial stages of breakdown.

It may be expected that all incorporation techniques will have some effect on diseases in the following crop. Appropriate defensive action with suitable sprays and seed treatments is at least a partial answer.

CULTIVATION – MACHINERY SYSTEMS

Incorporation and machinery choice

Ash incorporation
All that is necessary for ash incorporation is a very light harrowing. Depending on the soil and moisture content this can be achieved with a spiked harrow such as a zig-zag or with light discs.

Stubble incorporation
The machine that will give a completely clean seedbed is the mouldboard plough. However, chisels have been widely used in the UK for some years. Particularly on heavy soils, systems based on the chisel and chemicals such as paraquat have proved to be highly successful. Despite the talk in Germany about the reliance on the

mouldboard plough in that country, there are moves to move to chisel based systems to improve rates of work, reduce total power inputs and improve mixing.

Straw incorporation
The simplest view of the available technology and the safest is currently the mouldboard plough. However, particularly on heavy soils, there is likely to be a progressive move to chisels and discs. Powered cultivations may well be widely used but tend to be slow.

Machinery types

The mouldboard plough
The mouldboard plough is widely understood in the UK but there is not such wide experience of its use with incorporation systems.

On the continent, particularly in Germany, large clearances point to point and under-beam are widely recommended. For example, the continental norm would be 30 – 37 in under-beam while nearer 24 in would be common in the UK. It is also common in Germany to plough 30 – 35 cm (12 – 14 in) deep. ADAS work suggests 20 cm (8 in) will be adequate. The latest trials here also indicate that 30 cm (12 in) wide gives the best burial and mixing. The indications from ADAS work in the UK are that the conventional ploughs already on the farm are likely to be suitable for incorporation work.

The UK research suggests that the ideal situation is:

- a top layer, about 5 cm (2 in) deep free from trash, a layer of about 15 cm (6 in) in which the straw is thoroughly and uniformly mixed
- undisturbed soil below these layers, subsoiled if necessary.

It may be that shallow ploughs working at only 15 cm (6 in) deep could work adequately especially in shallow soils or with light crops of straw. If ploughing direct after the combine it looks as if it will be best not to cut the stubble. If pre-incorporation is to be carried out, then it will help to have the stubble short, chopped with a separate chopper if necessary.

Good skimmers and trashboards will be able to mix straw to some extent down the side of the furrow and still leave a trash free surface. It is important that the trashboard is adjustable to work at any depth at which the plough may work, including shallow work. Some of the continental ploughs have trashboards but they do not come into effect unless the body is used at the design depth of 30 cm (12 in) or more.

The general conclusion about ploughs is that it is best to use the sort of plough used locally and suitable for the soil, depth and conditions. All this can be within normal UK experience. The only real changes are likely to be trashboards for incorporating chopped straw. Even quite large quantities of straw, if it is well chopped as it needs to be, will flow through a normal plough without problems.

Plough presses can play a part in breaking up and consolidating on light soil behind the plough. They are useful at any weight in assisting in preparing the seedbed while the soil is still friable immediately behind the plough. Where there is straw to incorporate, extra weight will be helpful and it is desirable to get over 100 kg per wheel. Most presses so far available are significantly less than this. It may be best to use them 2 or 3 h or even days after ploughing to get maximum effect on heavy soils.

According to ADAS, there is no evidence that slatted mouldboards have any advantage over conventional full bodies. However, some farmers do feel they scour better on some soils at some moisture contents.

The problem with the plough is that it results in good yields on the sands, but the advantage decreases as the soil gets heavier. The disadvantages build up too. Draft on heavy soils may be two or three times as much as on light soils. Above 30 – 35% clay content the plough begins to run out of its usefulness. From 40% clay upwards, the chisel tends to be more attractive.

Chisels

While many people will turn to the plough to incorporate in the short run, in the long run many will look to chisels and discs to get higher rates of work and lower horse power hours per hectare. Even the Germans, widely held to revolve their systems round the mouldboard, are looking seriously at chisels.

German research

Koller and Stroppel carried out experiments with heavy cultivators at Hohenheim in Germany. Their main conclusions were as follows:

- For cultivators there needs to be a minimum clearance distance of 750 mm (30 in) between cultivator points and the base of the frame. The distance between tines in each direction should be about 750 mm (30 in) and the number and arrangement of tines should be chosen so that there is a distance of about 250 mm (10 in) between the cultivator furrows.

- The ground should be cultivated twice, the first operation at 100 – 150 mm (4 – 6 in) depth and the second deeper between 200 and 250 mm (8 and 10 in) depth.
- The spring tine cultivator did not maintain the set depth of work as well as the fixed tine cultivator.
- To achieve good incorporation and mixing throughout the depth of the profile, straw needs to be finely chopped. Experiments showed that material chopped to lengths between 0 and 50 mm (0 and 2.5 in) was mixed into a greater depth than material chopped in the range 0 – 100 mm (0 – 4 in).
- The greatest amount of straw decays when it is uniformly distributed into depths up to 200 mm (8 in). It is likely, therefore, that tines need to operate between 200 and 250 mm (8 and 10 in) so that the straw concentration may be evenly distributed.

Subsoil

We know that winter wheat or spring barley will push their roots down to around two metres in one season provided there is no impediment to root growth. If ploughing were the only cultivation, we would have to plough two metres deep. If the top 45 cm (18 in) is in good order, if the roots can get through that, they will go all the way on most soils. Subsoiling, then, is necessary on some soils under some conditions to break up pans and compaction to allow free root growth.

For this reason the chisel has been widely used by UK farmers. In fact, the UK leads the world in tine design. Work at the NIAE and Silsoe College have shown that a pattern with a 22.5 degree toe, a 45 degree foot and a vertical leg will give the least draught without bringing subsoil to the surface.

Discs

The main attraction of discs is rate of work. They can mix reasonably well, have very low maintenance requirements but may well compact the soil if worked repeatedly to the same depth.

The 'dish', or concavity of the disc will affect its power to penetrate, mix and invert. Deep concavity is used for primary cultivation with a high weight per disc. The Germans use 125 – 135 kg per disc for deep, primary cultivation. These large discs may well be up to 70 cm (28 in) in diameter. The sharper the angle to the line of travel, the greater the

Plate 3.13 The Parmiter Strawgon buries straw down a sub-soiler passageway.

penetration, mixing *and* draught. Tandem discs (using four sets with a hinge point in the middle) are easier to set than offset discs which leave a more level finish. (Offset discs are laid out with opposed front and rear sets, each mounted on a single axle. Sets in the UK will be up to 6 m wide.)

When you are looking at discs, examine for total weight, weight per disc and cost per lb. You will find they all cost within striking distance of £1 per lb weight. Try costing out alternatives then look at engineering quality. Look for taper roller bearings, silicone manganese steel discs, heavyweight chassis and well designed scrapers. They should be operated at 6 – 8 mph to get shatter and mix.

Powered cultivators

Early ADAS work showed that the rotavator was very successful at mixing chopped straw with soil. Lumkes from Holland has examined a wide range of existing equipment for incorporating chopped straw. His results showed that rotavation followed by two passes with a chisel plough gave the best mixing and incorporation with reasonable work rates.

The Chelli rotor spading machine has also been demonstrated to incorporate well, although it does not have a high rate of work.

A further option is to vibrate chisel tines using PTO power such as with the McConnel Shakaerator. The McConnel Tillaerator has proved to be remarkably capable in producing a seedbed in one pass. It will apply power direct to the soil without leaving smear on a pan. The power driven rear crumbler saves power and helps produce an even textured, even compacted, firm tilth. Because the tines push the tractor forward, this tool should ideally be coupled to a set of chisels to do the deep cracking and seedbed preparation in one pass. It can be coupled to a Shakaerator to get PTO power in use on both implements.

Combinations

With the many trials that have taken place in order to produce satisfactory incorporation, the disc/tine combination is most promising. The McConnel Discaerator was produced as a result of farmer demand. A set of discs is mounted in front of a 'Commando' cultivator

Plate 3.14 The Bomford 'Dyna-Drive' can be fitted with an incorporation screen to bury straw beneath a surface layer of straw.

and crumble Roller. It is a first rate tool to incorporate and deep crack the soil in one pass. For ash incorporation one pass will be adequate before drilling. For stubble and straw incorporation one pass will complete primary cultivation and mix straw in the top 15 cm (6 in). Subsequent passes with discs or harrow will be necessary to further move incorporated straw before final drilling. In fact a factory fitted option allows the chisels to be shaken by PTO power, i.e. they will be turned into a Shakaerator.

A medium weight set of cultivator tines can be used if required after the initial incorporation of straw followed by a pass of lighter equipment.

Another option is to put a set of zig-zag harrows behind a set of heavy discs. At the lightweight end of the equipment scale, discs, spring tines or light chisels can be coupled to a zig-zag harrow to produce a combination.

Drills

If the soil has been ploughed and there is a trash free seedbed, then conventional drills can be used.

If there is trash in the seedbed then the choice will be restricted. With limited amounts of trash, conventional disc coulter drills may work quite well. Such drills also work in scratch seedbeds after a burn.

Where larger amounts of trash exist in the surface, the drilling options are severely restricted. Current experience shows that the Moore All-till, the IH 511 and the Bomlet cultivation drill will all work with some success under such conditions. Drills using heavy press wheels built into their mechanism or followed by heavy Cambridge rolls are likely to ensure good seed/soil and straw/soil content. This will limit toxin problems and encourage rapid seed germination and establishment.

CULTIVATION ROTATION

There are arguments in favour of each type of cultivation system, each has its advantages and limitations or disadvantages. Mouldboard ploughing certainly has advantages in straw burial, but it is relatively slow, high in power demand and tends to produce a plough pan. Chisels mix the straw better, have higher rates of work, but leave a less attractive seedbed and may not control grass weeds as well. Each and every system can be described in this way. It follows, therefore,

Plate 3.15 The Moore drill can be used in tandem to give a 6m drill shown here working in a 'Nurse' stubble.

that the persistent use of any one system will accentuate the disadvantages. A periodic change from one system to another, a cultivation rotation, might, if properly managed, maximize the advantage of each system and minimize the disadvantages.

Many farmers have had good reason to look again at the mouldboard plough. With the grass weed problem we have – and sterile brome in particular – the plough is one of the useful weapons.

At the other extreme, direct drilling does appear to work very well for some. That system, though, can break down in the presence of couch and other grass weeds. Generally, chemicals have improved tremendously in the last ten years, but cereal growers still come up against weed problems.

The same sort of story applies to plant diseases. We have very much better control chemicals and procedures but there are still carry-over problems, which may get worse as straw burning in the fields is restricted.

For all these reasons and a few more, the original idea of crop rotation was developed. However, many of those reasons for crop rotation have now been controlled and the economic pressure on crop production is at least as bad and certainly different. It is, therefore, likely to be interesting, beneficial or necessary to change things around. Fifty years ago we had a constant cultivation system and rotated the crops. Now we tend to rotate crops less and cultivation rotation may provide the substitute we are looking for.

The ideal must be to go for low cost cultivation if it is possible. Logically, the best profit over the years must come from a system that aims at good yields from the lowest cost system that conditions will allow. If a cultivation system limits yield or by changing the cultivations you could help lift any other limit, then change the system; but only for long enough to remove the limit. So, the idea is to change the system as and when difficulties either arise or are foreseen. Since we operate in a living environment that is always changing we probably need to keep on changing the cultivation system.

The problem is to establish a set of rules that will indicate what to change and when. There are two questions: what to change and how much of your acreage. Insurance may be an important part of the system.

Figure 3.9 shows a cultivation rotation pattern. The idea is to move round the outer wheel of objectives using the inner wheel of methods. It is, of course, possible to draw up a different cultivation rotation using different methods or repetitions of patterns.

Straw incorporation

It may not be necessary to incorporate straw every year and, therefore, the method used can be rotated easily. If it is the mouldboard plough, that may also be the opportunity to use the same tool to control grass weeds by burial. Alternatively, if the incorporation system is to use chisels and discs, that may be the opportunity to subsoil with the chisels and use the discs to encourage weed seed germination followed by a kill with a herbicide such as Gramoxone. The systems of incorporation have been discussed above but how they fit into the overall pattern of cropping and the cultivation rotation is clearly important.

Compaction

Work has started on a series of trials at the Scottish Institute of Agricultural Engineering, SIAE. In the first year of a trial involving winter barley, yields went up by about 30% in a 'zero traffic' system

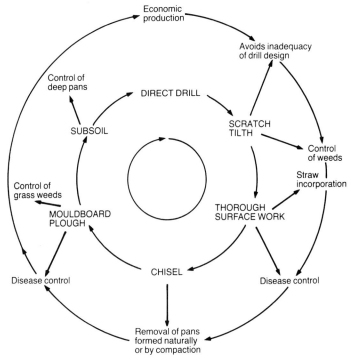

Fig. 3.9 Cultivation rotation.

compared with the same procedures on a normal system with temporary tramlines for fertilizer, top dressing and spraying. In the zero traffic system the wheels all ran on permanent tramlines outside the working width of the drill.

'In the conventional traffic system, all wheels ran over cropped areas of the plot with temporary tramlines (1.5 m track) for top-dressing and spraying. In both traffic systems, seedbeds were prepared by ploughing and secondary cultivation.

'Measurements in the seedbed showed that both traffic systems produced similar soil conditions to sowing depth, and similar plant populations. Below sowing depth, soil in the conventional traffic system had significantly greater bulk density, cone resistance and vane shear strength and lower air-filled porosity. In short, the soil was more compacted.

'Mean grain yields for the cropped areas exclusive of all tramlines were 6.5 t/ha for the zero traffic plots and 5.3 t/ha for the

Plate 3.16 Two passes with a good set of heavy discs will incorporate quite well.

conventional traffic plots. If the 12 m tramline system of the self-propelled gantries now becoming available had been employed, it is estimated that the overall field yields would have been 6.3 t/ha for the zero traffic system and 4.8 t/ha for the conventional traffic system. Observed crop growth and maturity and measured yield were much more variable within the conventional than the zero traffic plots.'

As the Head of the SIAE Soil Section, Mr Donald Campbell suggests, these results are not highly conclusive as they are from one trial in one year, but the situation was scientifically set up and carefully monitored. These results are indicative and highly suggestive that there really is something that can be done to increase crop yields by reducing compaction.

The results, say Mr Campbell, may be significantly altered by season and soil conditions but the differences would just as easily be larger as smaller. The 'zero' traffic system still contained significant wheelings, the point is that they were less than the tramline system. What Mr Campbell is convinced of is that we will go down this road of reduced wheelings using whatever methods, such as gantries or completely matched implements, that seem most practical.

Brian Finney, of ADAS, suggests that the right place to start is with the soil and a look at its past history. If it has been in minimal cultivation for some years, if there is a compaction it is likely to be above 12 in deep. If there is doubt, dig a hole, or, not as good but much faster, feel the soil with a soil auger. If compaction is there and below 12 in use a traditional subsoiler. If it is above 12 in use a Paraplow or a winged subsoiler such as the Haylock.

Tilth

Next, consider the end point; the seedbed and the crop. The seedbed is a key issue, evenness in all things is necessary to establish maximum use of plant food, soil water and the growing season. Evenness of depth of placement is high on the list and depends on an even depth of tilth and good coulter design on a drill driven not too fast.

The new generation of cultivators are at least interesting. The McConnel Tillaerator is PTO driven and does a thorough job but is rather slow; it can be linked to a subsoiler. The Bomfort Dynadrive has a low power consumption and is undoubtedly good under the

Plate 3.17 A standard leg Paraplow with mechanical trips.

right conditions. Probably most under-rated is the Taskers Tillage Train which demands a good tractor to pull it but does a good job at high speeds under a wide range of conditions.

These high speed cultivators are a substitute for a power harrow. Speed is a good clod breaker. In fact, discs have always been good at tilth formation. They work at high speeds, disturb all the land, have low energy inputs and give a good rate of work. Plain discs do, however, tend to compact. Tasker's new chisel discs should be a significant improvement.

Work at the SIAE in Scotland is going on concerning tilth preparation and seedbed conditions. Again, it seems, it is evenness that comes out as an important factor. Another under-rated tool is the Wilder Pressure Harrow on which the tines are retracted far enough to allow the harrow frame to run on the surface. The pressure system holds the frames on the surface to level it and cause even tine penetration. This even disturbance of the seedbed causing even consistency is the key to good coulter control on the following seed drill.

Weeds and diseases

Cultivation rotation is clearly a useful tool as a means of controlling weeds and diseases. The mouldboard plough may bury weeds and diseases and that may well be desirable at least initially. However, there may well be an element of preservation particularly with respect to disease. The Green Bridge, in particular, can provide a useful habitat for aphids if the green is ploughed in without being killed off first. Working straw and trash into the surface may have its attractions because breakdown can be speeded up and weed seeds germinated and killed off with subsequent cultivations or chemicals.

Insurance

Firstly, it is not likely that the same problems will occur over the whole farm at once. At least some of the time, problems are going to crop up on some fields and in some crops. So change may be necessary on some of the acreage and not on the rest.

It is also true that it will be possible to predict a problem which may crop up. For example, on NIAE test plots, yields from direct drilling were good and built up over several years but, after 6 – 8 years of the trials, grass weeds had built up and sterile brome in particular.

Something had to change and it was cultivations that were the solution.

Rotation conclusions

The implication is clear; all cultivation types have their long-term problems and must be changed from time to time as conditions demand.

There is also a weather and conditions insurance built into a system which uses several different systems on the same farm in the same year.

The conclusion must be that just as many move the subsoiler round the farm, so must the plough, the minimal seedbed and direct drilling be moved round. It may be advantageous from an insurance point of view, to mouldboard say 20% of each year, minimal cultivate say 40%, and direct drill or scratch seedbed the rest.

The precise rate of movement round the diagram will depend on the conditions, but, again, the aim must be to keep to the low energy systems.

COSTING AND DECISION MAKING

Costs and returns

Loss of yield as a result of the use of a particular technique can have serious effects on profitability. The trials at Boxworth up to 1983/4 showed marginal depressions in yield as a result of surface incorporation compared with burning. This depression was as high as 6 – 10% with some techniques in one year. Generally speaking, the yield depression following ploughing was less than that following surface, non-inversion incorporation.

The figures in Table 3.18 are taken from *Farm Management Pocketbook* by John Nix. It is clear that even a few percentage units drop in yield would affect gross margin enough to pay for significant extra costs in order to incorporate if such costs held yields up. The good manager, however, is the one who can keep yields up and costs down.

The easy answer is to look at the Boxworth figures and note that mouldboard ploughing used to incorporate could hold yield up at a level similar to a system based on burning and ploughing.

Table 3.18 Winter wheat returns

Production level Yield: t/ha	Low 4.9	Average 6.3	High 7.7
	£	£	£
Output: Feed wheat	540	695	845
Milling wheat	590	755	925
Variable costs:			
Seed		40	
Fertilizer		90	
Sprays		70	
Total variable costs per hectare		200	
Gross margin per hectare			
Feed wheat	340	495	645
Milling wheat	390	555	725

Plate 3.18 A 'trash-leg' version of the Paraplow suitable for working in incorporation seedbeds.

If we assume that the straw was incorporated before ploughing, it is fair to accept the whole cost of that ploughing as an extra cost. A typical contract charge for ploughing could be £35/ha and a farm cost of about £20 to £25/ha. Taking £25 as a guide, we could afford up to £25 loss of gross margin before ploughing became unprofitable. Looking at the figures in Table 3.21 above and looking at, say, average yields of feed wheat with a gross margin of £495, we could afford to lose about 5% of yield before ploughing became uneconomic. On the face of it, then, the Boxworth figures in Table 3.3 suggest it is worthwhile to plough to incorporate.

However, it may be remarked that plough based systems take more men, tractors and time than other systems. Maybe several times as much, as, say, the other extreme of direct-drilling. To mitigate against this, incorporation processes can start, should start, immediately behind the combine rather than waiting a week or two before a burn.

There are, of course, other pressing economic factors beside the immediate cost of ploughing. Table 3.19 shows the comparative costs of winter and spring cereals. Quite clearly a loss of winter cropping hectares due to lack of time to plough and consequential spring cropping is much greater than the likely yield depression in the autumn crop because of surface incorporation rather than ploughing.

This argument holds to a lesser extent if autumn planting is delayed. Lack of timeliness does have yield penalties. With winter wheat, for example, Fig. 3.10 shows that for every week sowing is delayed after 15th October, a yield loss of up to 200 kg/ha (1.6 cwt/acre) can be expected. In gross margin terms, that is £22 per hectare lost for every week of delay, i.e. abouut 6%.

So, while ploughing in chopped straw appears to be the number one choice, there are consequences which it may not be possible to accommodate every year without loss of profit overall.

Table 3.19 Gross margins of winter and spring cereals per ha (acre) (courtesy of John Nix)

Production level		Low	Average	High
Winter wheat	(feed)	340(137)	495(200)	645(262)
	(milling)	390(157)	555(225)	725(293)
Spring wheat	(feed)	225(91)	325(131)	425(171)
	(milling)	260(105)	370(149)	475(192)
Winter barley	(feed)	270(109)	400(162)	530(214)
	(malting)	320(130)	465(188)	610(247)
Spring barley	(feed)	245(99)	345(138)	440(178)
	(malting)	290 (117)	400(161)	505(204)

Alternative systems – costings

Figure 3.11 shows a comparison of costings drawn up by David Patterson of the NIAE in the late 1970s. The figures are dated but the balance and comparison has a relevance now. Generally, the higher output systems cost less except that the direct-drilling system used

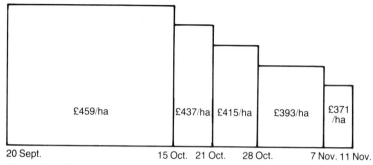

Fig. 3.10 Winter wheat gross margin.

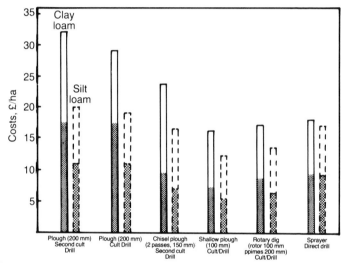

Fig. 3.11 Mean cultivation costs (1971-1977) at Boxworth (clay loam) and Rothhamsted (silt loam). The solid line denotes clay loam and the broken line silt loam. The dark and dotted areas (small dots) represent the cost of the primary cultivation on the clay loam and silt loam soils respectively. The unshaded area represents the cost of secondary cultivation and/or drilling. The dotted area (large dots) represents the cost of the spraying operation and the cost of paraquat. (courtesy of NIAE).

more chemicals which significantly added to the cost. There is no doubt that using chemicals to kill weeds does cost money.

Table 3.20 shows cost of tractors per operating hour on a full, real cost basis. Similarly Table 3.21, the full cost of a sprayer. Table 3.22 compares the machinery and chemical costs, on a full cost basis, of killing weeds by cultivating or sprayer. Table 3.24 makes a similar comparison based on marginal costings on money 'off the farm'. These figures can be further compared by contract charges as shown in Table 3.25.

Table 3.20 Cost of tractors

	76 hp Ford 5610	103hp Ford 7610
Basic data		
Basic list price	£12042	£14900
Discounted price with some extras	£12042	£14900
Fuel used per hour	2 gall	4 gall
Annual hours use	750	750
Annual cost of ownership		
Depreciation 13% of capital cost @ 13%*	£1565	£1937
Interest on capital, half av. cap. @ 10%†	993	1229
Repairs and maintenance @ 6.7%*	807	998
Tax and insurance	60	80
Total per year	3425	4244
Average per hour	4.57	5.66
Running costs per hour		
Man at £2.50 per hour + 10% (a)	2.75	2.75
Fuel and oil at 90 p/gall (b)	1.80	3.60
Total per hour	£9.12	£12.01
Rate of work acres/h	Spraying 14	Harrowing 7
Tractor cost per acre	0.65	1.72
Tractor cost per hectare	1.62	4.27
Money 'off farm' per hour (a) + (b)	4.55	6.35

*Method taken from *Farm Management Pocketbook* by John Nix.
†Method taken from *Farm Mechanisation for Profit* by Bill Butterworth and John Nix.

Table 3.21 Cost of implements

Type	Sprayer	Cultivator
	Vicon mounted 1100 litre 12 m boom	Konskilde Spring tine 6 m
Capital cost	£2600	£2140
Replacement life*	8	8
Average depreciation %*	8.5%	8.5%
Average depreciation cash	£ 221	£ 182
Spares and repairs		
% of capital @ 200 h use	9.5%	14%
Cash per year	£ 247	£ 300
Cash per hour	£ 1.23	£ 1.50
Total cost per year	£ 468	£ 482
Cost per hour @ 200 h	£ 2.34	£ 2.41
Rate of work/hour acres	14	7
Implement cost per acre	0.17	0.34
Implement cost per hectare	0.42	0.85

Note: The 'spares and repairs' percentage allowances are taken from John Nix's *Farm Management Pocketbook.* Clearly, the figure for cultivators will vary considerably depending on soil type, depth of work and so on. This figure could easily be much higher on many soils.

Table 3.22 Summary of machinery costs

	Sprayer	Cultivator
	£	£
Tractor cost/acre	0.65	1.72
/ha	1.62	4.27
Implement cost/acre	0.17	0.34
/ha	0.42	0.85
Total cost/acre	0.82	2.06
/ha	2.04	5.12
Chemical cost per acre	5.09	
@ 2 l/ha per ha	11.94	
Total machine and chemical costs		
per acre	5.91	2.06
per ha	13.98*	5.12

* discounts on chemical could knock £2/ha off these figures.

Table 3.23 Costings of spraying v. cultivating (costs/ha) *(Farm Management Pocketbook,* John Nix 1985)

	Rate of work ha/8 h day		Contract charge	Average farmer's cost
	Av.	Premium		
Spring tine harrowing	10	16	14.00	6.35
Spraying (medium volume)	12	20	14.00	5.50
(high volume)	10	15	24.00	12.50
Chemical			11.94	11.94

Table 3.24 Costings of spraying v. cultivating ('money off the farm' or marginal costings; extra cash spent to do the job)

	Spraying	*Cultivating*
Tractor running costs/h (from Table 3.20)	4.55	6.35
Cash from Table 3.21		
Running costs per hour, spares and repairs	1.23	1.50
Total (a)	5.78	7.85
Rate of work acres/h (b)	14	7
Marginal machinery		
Cost per acre (a) ÷ (b)	0.41	1.12
ha	1.01	2.77
Cost of chemical per acre	5.09	
per ha	11.94	
Total 'money off farm' acre	5.50	1.12
ha	12.95	2.77

Cost effectiveness

Weed competition

Clearly spraying is significantly more expensive than cultivating. However, there is a question of cost effectiveness. As Tables 3.15 and 3.16 quite clearly show, cultivation may not kill off weeds, particularly grass weeds. Cultivation may even make matters worse. If soil conditions are damp before or within several days after cultivation, weeds may be just transplanted rather than killed. Weed competition does affect crop yield, possibly by several percent.

Effectiveness of operations WRO experiments 1979:

	Heads/m^2 July
Paraquat + cultivation	364
Cultivation only	882

Table 3.25 Cost-effect balance

	Sprayer	*Cultivator*
	£	£
Cost to cultivate per ha (from table above)		5.12
Value of 1% loss in yield		10.50
(for 10 tonne crop at £105 per tonne)		
Total cost of cultivating and loss of		
1% yield		15.62
2% yield		26.12
5% yield		57.62
10% yield		110.12
Cost of spraying/ha	13.98	

It is, in fact possible to put a guide line figure. Steve Moss (of the WRO) says that for blackgrass (Alcopecurus myosuroides) populations between 0 and 200 plant m^{-2}, a good guideline is 1 tonne lost yield per ha, per 100 blackgrass plants per square metre. These numbers are probably best counted in February or March but the rule is likely also to hold for autumn counts.

It is further known, from WRO data, (Tables 3.18 and 3.19) that transplanting of blackgrass by cultivation in damp conditions can result in multiplying up the strength of the plant and maybe the number of heads by a factor of four. Under such circumstances 25 plants of blackgrass m^{-2}, says Mr Moss, could lose a tonne yield/ha. This sort of loss has happened in trials, and presumably, can and does happen in the field. This sort of figure, also, according to the research, applies to wild oats and other weeds.

The implication is quite clear. One tonne loss of yield is in the region of 10 – 15%. It could be less or even significantly more. Clearly, weed control measures are not enough if they do not achieve their objective. Even quite small populations of weed, resulting from inadequate control, will result in significant loss in yield.

Plate 3.19 Destroying the 'green bridge' on stubble needs to be effective. Spraying is not initially the cheapest way of doing the job but it will normally be the most cost effective.

Soil structure

Soil surface conditions may be improved by cultivation. On the other hand, on silts or low organic clays, surface capping can easily result from over cultivation of the soil surface. Again this can significantly affect crop yield.

Water retention

Cultivation will, in the dry, inevitably result in some water loss. In dryer seasons or in the dryer parts of the country this may significantly affect crop establishment and final yield.

Costs/yields

A 10 tonne/ha yield of wheat at, say £105 per tonne would be worth £1050 per ha. A 1% loss in yield would be worth £10.50 per ha.

The difference in costs of spraying and cultivation to control weeds, from Table 3.22 would be £8.86 per ha (or, say, £6.86 with discount).

Costings can, of course, be worked out in different ways. 'Money off the farm' is shown in Table 3.24 and contractors' charges in Table 3.23. Such figures do show marginal charges in the difference in cost between cultivating and spraying. The differences however, are still very small if there is any loss in yield of the crops due to inadequate weed control.

A 1% loss in yield, therefore, due to inadequate weed control, poor soil surface conditions and/or loss of moisture from the seedbed, would more than cancel out the difference in cost between the methods of control. As the research shows, 10 – 15% less yield is quite possible and certainly does often occur. Cultivation *can* be more cost effective but it must actually result in killing weeds if it is not to result in yield penalties. *In practice*, spraying is likely to be a faster, surer method of control without complications or possible losses.

Conclusion

There is clearly a dilemma in weed control before planting. Spraying is more expensive than cultivating. The choice depends on an assessment of the conditions at the time. Weather, soil conditions and annual rainfall are all involved and what is being looked for is an 'insurance policy'. Cultivation is initially cheaper. If, however, it is not carried out to be fully effective, even quite small extra weed populations will easily affect yield enough to cancel out any cost advantage.

Weed control – the main chemical options

Burning will destroy some weed seeds and encourage others to germinate. Ash incorporation will also tend, by mechanical abrasion and seed incorporation, to encourage that germination.

Paraquat

Various formulations and trade names for paraquat will have marginally different effects. The ICI recommendation for paraquat as Gramoxone is 5.5. l/ha for crops direct-drilled into stubble. This may rise to 8.5 l/ha for grass destruction or be as low as 3 l/ha for annual weeds in an arable seedbed. Paraquat is good for control of annual weeds including broad leaved species. Timing of paraquat is not a problem provided a reasonable area of leaf is exposed. It is best to wait for grass weeds and volunteer cereals to have at least one leaf fully expanded. The major advantage of paraquat is being relatively rain-fast. It can be used in 'catchy' weather which so often happens in a British autumn. List price (Jan. 1985) of Gramoxone at the recommended rate of 2 l/ha was £11.94 per ha.

Glyphosate

The major attraction of glyphosate is its translocation. At high rates of 4 l/ha it will provide good control of couch. At the low rate of 1.5 l/ha it has good control of grass weeds, such as blackgrass, wild oats and volunteer cereals even with only partially expanded leaves. The high rate is, however, needed for good control of broad leaved weeds. List price (Jan. 1985) of glyphosate at the recommended low rate of 1.5 l/ha was £22.95 per ha.

Alternative systems – planning the decision

Gramoxone, for example, has a high degree of rain fastness. Cultivating on damp conditions will not only not kill, the transplanting effect has been shown at WRO to actually increase weed populations, particularly blackgrass, under some conditions. Secondly, spraying is likely to be somewhere near twice as fast as cultivating, possibly more. Thirdly, soil surface conditions and moisture retention have to be brought into the equation.

There is a question then about not just the technique to be used, but how it fits in with the farm as a whole, its soil type, its resources and the time available in each year. The concept of cultivation rotation (p. 86) is, in practice, an important management tool.

Plate 3.20 Howards combination of tines, Rotavator and crumble roller can do an excellent job in one pass.

Burn, stubble or chop
On some farms, on some fields, in some years, burning is likely to remain a feasible possibility. Particularly on heavy soils where ploughing tends to limit yields this will remain an attractive option. Strict compliance with the Straw Burning Codes, the Law and good local PR informing people about the work, may help preserve this option as a practical possibility.

Burn
After a burn, ash incorporation will be obligatory. On light soils ploughing may well be the best option from a yield point of view. Scratch tillage followed by herbicide treatment will be very fast and the best yield option on heavy soils. In such a case disc or tine drills such as the Moore, IH 511 or Bamlett will probably give the best results if there is any trash about.

Stubbles
Where stubble cleaning and minimal tillage without mouldboard ploughing have been practised, the same practice can be followed, nothing will have changed. High rates of work with sensible chemical use can help produce clean crops planted at the optimum time.

It may well be that the idea of Nurse Stubbles using suitable drills to plant direct into a harrowed stubble can be developed. There is already a little experience of this technique and it may be possible to achieve good yields of both autumn and spring crops following very high rates of work.

Straw incorporation
The safe option for many looks like being the mouldboard plough. On heavy soils we will be immediately looking at tines and discs and, in the long run, this possibility is likely to be developed for the lighter soils.

The decision

Table 3.26 will help summarize the present situation, the targets and the judgment about which technique to use based on the arguments in pages 34 – 65. Table 3.26 indicates the choice of each implement and how it should be worked.

Added to this has to be a plan integrating the procedure and the whole farm and the detailed husbandry for each field.

Conclusion
On many farms in many years, it will be prudent to use more than one technique to manage straw. These techniques may change from year to year. Overall best profit will come from matching the techniques used to the resources and time available, balancing costs and returns.

Table 3.27 Cultivation targets

	Yes/No Objective	*Minimum depth*		*Preferred implement*
1.	Ash incorporation	5 cm	2 in	Discs or tines
2.	Stubble incorporation	7 – 8 cm	3 in	Discs/tines or plough
3.	Straw incorporation	15 cm	6 in	plough or discs/tines
4.	Subsoiling	60 cm	24 in	Subsoiler tines
5.	Deep cultivation	35 – 40 cm	14 – 16 in	Chisel tines
6.	Destruction of Green Bridge	5 cm	2 in	Tines or sprayer
7.	Trash or weed burial	15 cm	6 in	Mouldboard plough

Table 3.26
Farm details

Address: ....

Sheet filled in by:..

Total acres cereals...

Total acres burned (for ash incorporation)..

     acres baled (for stubble incorporation) ....

     acres chopped (for full straw incorporation)..

Soil type:...

Soil condition when action required ....

Possible depth of working at time relevant ....

Implements available: Max. working depth

1..   ....

2..   ....

3..   ....

4..   ....

5..   ....

Tractors available: Horse power:

1..   ....

2..   ....

3..   ....

4..   ....

5..   ....

Preferred implement, procedure and timing:

1 (Ash) ....

2 (Stubble)..

3 (Chopped straw) ....

Fall back procedure:

1 (Ash) ....

2 (Stubble)..

3 (Chopped straw) ....

LIGHT SOIL

Options

1. Burn where feasible and desirable;
2. nurse stubble drilling after baling;
3. incorporation of stubble or full straw crop.

PREFERRED INCORPORATION OPTION. ACTION PLAN 1.

Plan 1 notes:

1. Generally, if ploughing, chopping on the combine is likely to be most attractive. The chop length should, in practice, be short enough to allow easy passage through the plough. Fine chopping speeds breakdown but costs energy. Reasonable target is 90% in the 5 – 7 cm (2 – 3 in) range.

2. and 3. There appears little reason to delay and from a timeliness point of view the plough could go in behind the combine to allow maximum time for the straw to be in contact with the soil before planting.

4. On light soils a really heavy plough press can be coupled to the plough. It is possible to destroy weeds in the "stale" seedbed later by spraying. This can then be followed with the drill.

5. Destruction of weeds reduces competition and disease carry-over. Destruction by cultivation will not be successful unless there are several days of dry weather before and after cultivation. Even then, it is unlikely to be as effective as spray which is also twice as fast. Maintenance of seedbed moisture should also be considered.

5c. and 6. Cultivation does affect surface conditions. Over cultivation will have adverse effects on the establishing seedling and crop yield. If using cultivation to control weeds, this should also produce the seedbed. If spraying to achieve a clean seedbed, cultivations can be kept to a minimum.

7. The drill can easily be any conventional drill to be chosen or already in service. Look for good depth control and keep speeds low to maintain this evenness of depth of planting.

DECISION PLANNER — LIGHT SOIL
Plan 1: Preferred option

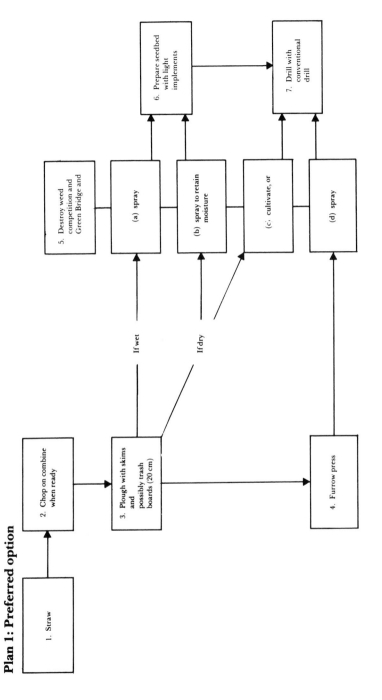

LIGHT SOIL

ALTERNATIVE INCORPORATION OPTION. ACTION PLAN 2.

Plan 2 notes:

1. Generally, if ploughing, chopping on the combine is likely to be most attractive. The chop length should, in practice, be short enough to allow easy passage through the plough. Fine chopping speeds breakdown but costs energy. Reasonable target is 90% in the 5 – 7 cm (2 – 3 in) range.

2. Combine chopping has organizational and energy advantages. However, a long stubble attached to the root ball is very difficult to incorporate. If combine chopper, cut stubble short. If long stubble, combine output increases but use full width stubble chopping. Stubble chopping is easier and gives more even distribution if straw is spread by the combine. Swath spreading is a further compromise that takes less time than stubble chopping but leaves some of the stubble/root ball problem. Final chop including the stubble should give 90% in the 5 – 7 cm (2 – 3 in) range.

3. Cultivation with straight legged, rigid tines appears to be the best coupled to the use of discs. The minimum depth should be to 15 cm (6 in) preferably 20 cm (8 in) with heavy crops of straw.

4. Destruction of weeds reduces competition and disease carry-over. Destruction by cultivation will not be successful unless there are several days of dry weather before and after cultivation. Even then, it is unlikely to be as effective as spray which is also twice as fast. Maintenance of seedbed moisture should also be considered.

4c. and 5. Cultivation does affect surface conditions. Over cultivation will have adverse effects on the establishing seedling and crop yield. If using cultivation to control weeds, this should also produce the seedbed. If spraying to achieve a clean seedbed, cultivations can be kept to a minimum.

6. The drill can easily be any conventional drill to be chosen or already in service. Look for good depth control and keep speeds low to maintain this evenness of depth of planting.

DECISION PLANNER — LIGHT SOIL

Plan 2. Alternative for speed and low energy inputs (possibly yield penalty)

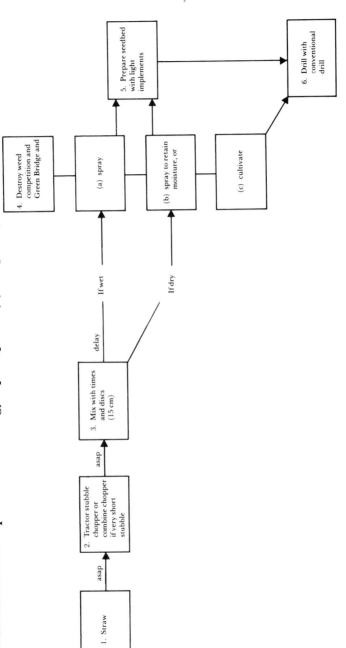

HEAVY SOIL

Options

1. Burn where feasible and desirable;
2. nurse stubble drilling after baling;
3. incorporation of stubble or full straw chop.

PREFERRED INCORPORATION OPTION. ACTION PLAN 1 (all but heaviest soils when wet)

Plan 1 notes:

1. Generally, if ploughing, chopping on the combine is likely to be most attractive. The chop length should, in practice, be short enough to allow easy passage through the plough. Fine chopping speeds breakdown but costs energy. Reasonable target is 90% in the 5 – 7 cm (2 – 3 in) range.

2. and 3. There appears little reason to delay and from a timeliness point of view the plough could go in behind the combine to allow maximum time for the straw to be in contact with the soil before planting.

4. Destruction of weeds reduces competition and disease carry-over. Destruction by cultivation will not be successful unless there are several days of dry weather before and after cultivation. Even then, it is unlikely to be as effective as spray which is also twice as fast. Maintenance of seedbed moisture should also be considered.

4c. and 5. Cultivation does affect surface conditions. Over cultivation will have adverse effects on the establishing seedling and crop yield. If using cultivation to control weeds, this should also produce the seedbed. If spraying to achieve a clean seedbed, cultivation can be kept to a minimum.

6. Where extra work has to be done on the tilth, there may be an argument to retain moisture by doing the final seedbed cultivation close to or actually with the drill. Hence the use of the combination of power harrow and drill. Conventional drill is adequate. Keep drill speeds low to maintain accuracy of depth of placement.

DECISION PLANNER — HEAVY SOIL

Plan 1. Preferred option

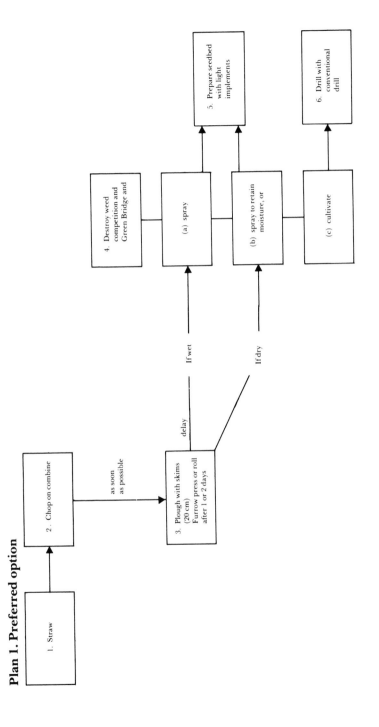

1. Straw

2. Chop on combine

as soon as possible

3. Plough with skims (20 cm) Furrow press or roll after 1 or 2 days

delay

If wet

If dry

4. Destroy weed competition and Green Bridge and

(a) spray

(b) spray to retain moisture, or

(c) cultivate

5. Prepare seedbed with light implements

6. Drill with conventional drill

HEAVY SOIL

ALTERNATIVE INCORPORATION SYSTEM. ACTION PLAN 2

Plan 2 notes:

1. and 2.   Generally, if ploughing, chopping on the combine is likely to be most attractive. The chop length should, in practice, be short enough to allow easy passage through the plough. Fine chopping speeds breakdown but costs energy. Reasonable target is 90% in the 5 – 7 cm (2 – 3 in) range.

3.   Pre-incorporation to a shallow depth only will help avoid a layer of straw at plough depth. There is some evidence that this can be a problem as the crop develops. Tines and/or discs, possibly in combination, should be in behind the combine as soon as possible.

4.   Ploughing to 20 cm (8 in) is required to spread the straw through the soil. Deeper is not necessary. The furrow press direct on the plough may smear and compaction of the furrow slices may be better hours or even days later using a very heavy press.

5.   Destruction of weeds reduces competition and disease carry-over. Destruction by cultivation will not be successful unless there are several days of dry weather before and after cultivation. Even then, it is unlikely to be as effective as spray which is also twice as fast. Maintenance of seedbed moisture should also be considered.

5c and 6.   Cultivation does affect surface conditions. Over cultivation will have adverse effects on the establishing seedling and crop yield. If using cultivation to control weeds, this should also produce the seedbed. If spraying to achieve a clean seedbed, cultivations can be kept to a minimum.

7.   Where extra work has to be on the tilth, there may be an argument to retain moisture by doing the final seedbed cultivation close to or actually with the drill. Hence the use of the combination of power harrow and drill. Conventional drill is adequate. Keep drill speeds low to maintain accuracy of depth of placement.

DECISION PLANNER — HEAVY SOIL

Plan 2. First alternative if heavy and wet over winter, i.e. anaerobic conditions

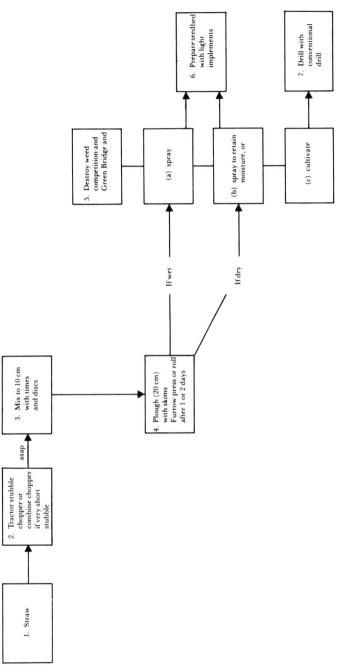

HEAVY SOIL

SECOND ALTERNATIVE INCORPORATION SYSTEM. ACTION PLAN. Heavy soils and low energy/high speed systems.

Plan 5 notes:

1. Generally, if ploughing, chopping on the combine is likely to be most attractive. The chop length should, in practice, be short enough to allow easy passage through the plough. Fine chopping speeds breakdown but costs energy. Reasonable target is 90% in the 5 – 7 cm (2 – 3 in) range.

2. Combine chopping has organizational and energy advantages. However, a long stubble attached to the root ball is very difficult to incorporate. If combine chopper, cut stubble short. If long stubble, combine output increases but use full width stubble chopping. Stubble chopping is easier and gives more even distribution if straw is spread by the combine. Swath spreading is a further compromise that takes less time than stubble chopping but leaves some of the stubble/root ball problem. Final chop including the stubble should give 90% in the 5 – 7 cm (2 – 3 in) range.

3. The minimum depth for straw distribution appears to be at least 15 cm (6 in). The information is that in heavy crops, 20 cm is a more reasonable target and this will only be so with reasonable consistent fine chopping.

4. Destruction of weeds reduces competition and disease carry-over. Destruction by cultivation will not be successful unless there are several days of dry weather before and after cultivation. Even then, it is unlikely to be as effective as spray which is also twice as fast. Maintenance of seedbed moisture should also be considered.

4c. and 5. Cultivation does affect surface conditions. Over cultivation will have adverse effects on the establishing seedling and crop yield. If using cultivation to control weeds, this should also produce the seedbed. If spraying to achieve a clean seedbed, cultivations can be kept to a minimum.

6. Where extra work has to be done on the tilth, there may be an argument to retain moisture by doing the final seedbed cultivation close to or actually with the drill. Hence the use of the combination of power harrow and drill. Conventional drills may not work well in a trashy seedbed; use trash drill such as Moore, Bamlett or IH.

DECISION PLANNER — HEAVY SOIL

Plan 3. Second alternative
— very heavy soils where ploughing not easy or unsuitable;
— cases where low time and energy inputs are priorities.

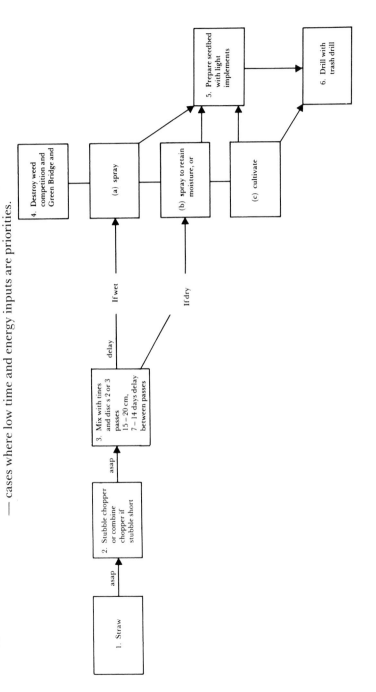

4

Harvesting, handling and storage

HARVESTING

It was clear that the advent of the combine following the stationary thresher resulted in a straw that was comparatively bent and bruised. It was unsuitable for many traditional applications such as thatching. The advent of rotary combines has produced a similar jump in the condition of straw; many varieties break up into shorter lengths and so much so that it is difficult for some types of baler pick-up to gather the swath. Many types of rotary baler find it difficult to hold these short lengths without significant losses back onto the ground.

Straw spreading behind the combine is desirable before a burn or in front of a stubble chopper. Spreading before chopping reduces peak power demand, improves rates of work and helps produce an even spread of chopped straw. Power demand from the combine is low – normally well under 7.5 kW (10 hp).

Chopping on the combine will demand more power, possibly 30 kW (40 hp) or more. Under conditions with damp or wet straw this may have a significant effect on the performance of the combine.

The larger combine in heavy crops will produce a swath which is also large. This may have problems going under the tractor when the baler is in-line as is the case with large balers.

The key factor in collecting straw is that it is a high bulk/low density and low value product. It is rarely possible to load up a tractor to the limit of its capacity. The key to high rates of work and

Plate 4.1 The flat-eight system is by far the most common but flat-ten will handle 40% more straw at one bite.

low costs is, then, to keep tractor inputs down. Tractor costs are much higher in real terms than many farmers easily accept. Even the marginal costs of labour and fuel are significant. Depreciation adds substantially to these figures.

Table 4.1 shows tractor costings including labour. A tractor for today's high speed, high density baling would have a capital cost of probably between £15 000 and £20 000. Its total running cost, therefore, would be in the region of £10 – £15 per hour if full depreciation at today's values were accounted. Table 4.2 shows the cost of running a baler for 200 h per year. Table 4.3 shows the cost of running a simple large bale handling system. In the case of that admittedly low cost system, tractor and man power would cost over £10 per hour by any reasonable measure and the bale handling equipment under £1. There is at least a factor of 10 and probably even 20 involved. So, anything that can be done to improve the productivity of the man on the tractor is worth looking at.

Table 4.1 Costings of three tractors at a range of tractor hours

	Typical small tractor costing £8000			Medium/large tractor taken at £20 000			Large tractor taken at £30 000		
Annual hours' use	300	800	1 200	300	800	1 200	300	800	1 200
Life in years	20	15	12	20	15	12	20	15	12
Repairs and maintenance as percentage of capital cost	4%	7%	9%	4%	7%	9%	4%	7%	9%
Annual cost of ownership (£):									
Depreciation (capital)/(life)	400	533	666	1 000	1 333	1 666	1 500	2 000	2 500
Interest (half at 10%)	400	400	400	1 000	1 000	1 000	1 500	1 500	1 500
Repairs and maintenance	320	560	720	800	1 400	1 800	1 200	2 100	2 700
Tax and insurance	50	50	50	130	130	130	200	200	200
Additional cost:									
Man £3.00/h plus 10 h/100 tractor hours	990	2 640	3 960	990	2 640	3 960	990	2 640	3 960
Fuel and oil at £1/gall	600	800	2 400	1 800	4 800	7 200	3 000	8 000	12 000
Totals/year	2 760	4 183	8 196	5 720	11 303	15 756	8 390	16 440	22 860
Cost/engine hour	9.2	5.23	6.83	19.07	14.13	13.13	27.97	20.55	19.05
Fuel at respectively	2 gall/h			6 gall/h			10 gall/h		

Table 4.2 Costings of an expensive large round baler, valued at £9000. Based on 200 h annual use

Capital cost	£9000
Life in years	9
Cost of ownership:	
Depreciation	£1000
Interest on capital (half at 10%)	£ 450
Repairs at % capital cost	7%
Repairs cost (String not included)	£ 630
Total/year	£2080
Cost per hour	£ 10.4

Table 4.3 Bale handling equipment

Equipment:	
Fork attachment for foreloader and 3 point prongs to pick up 2 at the rear. Total cost say £600	
Depreciation	£60
Interest on capital	£30
Repairs at 7%	£42
Total/year	£132
Cost per hour	£ 0.66

Plate 4.2 A flat-ten bale sledge; note the five channels which will help the bales bind in the stack.

Bale density

Bale densities have improved steadily in the last ten years and are probably 50% up on average. Good bale density reduces the 'dead' time when a large round baler is ejecting its bale because there are fewer bales. Fewer bales also means lower string costs and less handling. Improving density by 50% will cut the number of bales, and hence handling costs by 33.3%. Dense bales handle, stack and weather better. They also demand less storage space.

More bales per load

Clearly, more bales of whatever size, or shape, built into each tractor load will improve the efficiency of the man and his tractor.

BALING: METHODS AND COSTS

The written down annual cost of ownership of a baler, plus its maintenance costs are compared with the tractor and labour costs outlined above (unless a self-propelled collecting system is employed).

The man and tractor will have a real running cost far in excess of the cost of the rest of the system (i.e. the sledge or other handling equipment). Within reason, therefore, attachments or equipment which speed up the job are worth consideration.

Figure 4.1 shows a comparison of some systems of handling bales based on ADAS survey work published in 1975. Clearly the flat eight systems which are the most popular mechanized system in the UK have an average performance. The later conventional bale handling systems such as the flat-ten can however, compete with the large bale systems in terms of man/min per ton handled at least.

Eventually all handling systems have to be compared on a basis of cost. There are many practical considerations involved in bale handling but the most difficult to come to terms with is cost and this is for two reasons. Firstly, it is difficult to know what the true cost of the men, tractors and other resources really is because they tend to be spread over the whole farm for the entire year and, in any case, the involvement varies from year to year. Secondly, the nature of this high volume/low value material makes cost highly critical.

The costings shown in Table 4.4 show quite clearly that a flat eight system that takes only eight bales at one run has serious limitations, regardless of the cost of running the man and tractor.

Plate 4.3 There have been various systems, like the Lely, for handling mini stacks of conventional bales.

Systems

The systems in Table 4.4 outline a range of the possibilities available. The flat-ten is the complete handling process including the trailer. The double flat-eight process includes an accumulator that puts one flat-eight on top of another so that the loader fork picks up 16 bales.

The Freeman process from America uses a self-propelled unit that has a fork-lift truck built into it. The complete machine costs about £60 000.

The large bale mover is an automatic trailer with a chain in its bed so that it can self-load large bales on the move.

The automatic trailer is one of the Sperry New Holland range for conventional bales.

A comparison of the processes shows that the best types are those that aim at picking up a reasonable quantity of bales in one run. This argument applies whatever tractor and man costs are applied. Even when the tractor is not costed into the baling at a realistic rate, the result still comes out in favour of handling as much as the tractor can

Assume: 1. Transport distance 900m (1000 yd) each way
2. Trailer unaccompanied
3. 150 bales per load
4. Transport speed 2.2 m/s (5 mph)

1. Mounted bale carrier: hand built Heaps carrier elevator hand stack

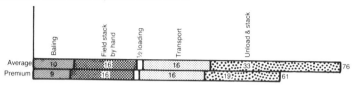

2. Squeeze loader 4 × 2: hand built Heaps loader trailer elevator hand stack

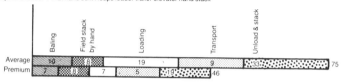

3. Flat 8 accumulator: impaler loader trailer stack with loader

4. Flat 10 accumulator: impaler loader trailed carrier stack direct from carrier

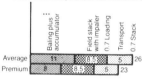

5. Big bale: gripper trailer stack with gripper

KEY

Baling

Field stacking/arranging

Loading

Transport

Unload & stack

5   10   20   30   40   50   60   70   80

man min/tonne

Fig. 4.1 Comparison of bale handling systems (*Source:* ADAS 1975).

Table 4.4 Summary of system costs including baling and handling (Index of cost per tonne at various costs for 200 h annual use of the basic unit)

System	Tonnes handled	Index with Flat 8 taken as 100
Flat-ten	1088	82
Double flat-eight	608	66
Freeman unit	2438	61
Accumulator fork	263	125
Large rectangular bale (carrying 1 bale per run)	1090	80
Large bale mover (to field rows only)	968	60
Flat-eight	800	100
Cube-eight (carrying 48 bales, eight per run)	800	70
Strapped package	840	93
Automatic trailer*	1500	75
Large round baler (5 ft) (carrying 1 bale per run)	968	83
Large round baler (4 ft) (carrying 1 bale per run)	713	109

* Conventional bales
Figures first published in Farmer's Weekly in 1977 resulting from field research by W.R. Butterworth. Figures here indexed to account for inflation. Main actual cost is man and tractor.

Perhaps the most significant measure of performance of a bale handling system is hidden in this table. The table was based on 200 h use of a system per year.

manage at one run. In fact, when the calculations are done again with, say, different string prices, fuel costs or any other variable, the result is the same. The capital costs of each process should also be noted.

The table shows the figures for 200 h annual use only, on the grounds that most farmers tend to say they have about that amount of time available each season to do the job, and then choose a process that will do it within that time. Obviously in 200 h, a different process will handle a different amount.

When capital is difficult to find an accumulator fork has obvious attractions.

At the other extreme, where the scale is available, the high capital cost of a process like the Freeman at maybe £60 000 may be highly competitive, because of its massive capacity. At higher levels of annual use, the high capital cost processes have even greater advantages.

The secret is to load up the tractor and use the driver's time without waste. A 45 kW tractor picking up only 150 kg at a time is wasting useful potential.

Plate 4.4 The Farmhand Forcalator produced a flat-eight package without using a sledge. Capital costs were reduced with only marginal effects on rate of work.

Flat-eight

According to ADAS figures, an average workrate to cover a transport distance of 900 m (1000 yards) is, in practice, 18 min for loading, 5 for transport and 11 for stacking – a total of 34 man minutes per tonne.

Baling rate was recorded at an average of 15 man minutes per tonne. So, for every hour spent baling and accumulating, 136 min or 2.26 h were spent carting and stacking. Consistently good operators with the windrowing sledge would improve on these workrates by 40% and consequently reduce costs.

Flat-ten

Why do some farmers buy handling processes other than the popular flat-eight? Some prefer the bigger machine – the flat-ten. It has several simple advantages. It has two more bales in each 'bite' and the bales are larger.

Plate 4.5 The McConnel Balepacker produced the advantages of big bale package handling while retaining small bales for easy feeding out.

Instead of being based on a 36 in imperial standard bale (just under 1 m), it is based on a 48 in bale (1220 mm). The combination of larger bales and more of them can increase the weight of the bite by 40%. Using a bit more of the massive margin of tractor power available makes better materials handling sense.

The flat-ten make-up also means that, when stacked at right angles to each other, the layers of bales 'bind' in the stack. A stack on a flat-ten base is much more stable – and can stand being moved several times. It is possible to pick up the stack mechanically in, say, 100 bale lots, and transport it by purpose-built trailer. One man can then handle reasonable tonnages at a fairly high speed without leaving the tractor seat. In fact, ADAS surveys, which compared the flat-ten process, including the automatic trailer, with the large square bale system, showed that, under a wide range of conditions, the flat-ten could perform very competitively and was slightly faster.

The binding-in-the-stack feature also allows the loader fork to be used for stacking and unstacking, although if the mechanical de-stacking is to be successful, the stack has to be constructed

Plate 4.6 The latest generation of large round baler has very much higher density and much more control over the bale.

carefully with well-packed bales in good condition.

A disadvantage of the flat-ten is that it is initially expensive compared with other methods, and some of the early machines put on the market had considerable teething troubles. However, not all farmers bought the flat-ten complete, including the trailer.

Double flat-eight
Another alternative to the flat-eight is basically the same. It consists of two flat-eights put together so that 16 bales are lifted at a time. This idea was originally put forward by Fossway Engineering, Midsomer Norton, near Avon, who designed and built it. The firm substituted a loader clamp for the flat-eight fork – and it worked.

Accumulator fork
For the accumulator fork handling process, an accumulator is built into the loader fork to produce a flat-eight system which has a low capital cost. The accumulator fork is just that, but the driver's workrate is slightly reduced. The machine, however, frees the baler

Plate 4.7 The Howard square bale was the forerunner in the UK of a new era in handling.

from any encumbrance behind and, despite the manufacturer's claims to the contrary, the ADAS surveys show clearly that implements hung on the back do slow the process down. For this reason, there was some movement away from accumulators hung on the baler itself.

The accumulator fork is used with wheels and a 'guiding frame'. The operator drops the fork on the ground wheels, aims his tractor at a line of single bales dropped by the baler, and puts two bales down at each of four entrance gates.

When full, the machine is used exactly like any other flat-eight fork to carry, load and stack. Because the operator is covering only slightly more ground than he would to clear the same field if collecting from pre-formed flat-eights, the theory is that he should be able to work nearly as fast with the added attraction of low capital cost and an uncluttered baler operation.

Four × two accumulator
The 4 × 2 accumulator method provides an eight-bale stack, built four bales high, in layers of two bales each. From a weathering point of view, it is an attractive handling unit. Bales are left as the manned

Plate 4.8 The Hesston square bale has a very high density and is very convenient to handle but the baler itself is very expensive in the UK.

sledge used to leave them, and it has attractions for the field curing of hay.

Because the stack is only two bales wide at the base, it is possible to conveniently pick up several stacks at once with machines that are not too cumbersome. This system was brought in by British Lely but never really caught on.

Strapped package

Another bale handling method is to put conventional bales into a larger pack and strap them up. This is claimed to achieve the advantages of conventional and large bales. Although mechanically it is difficult to achieve this target, for some operators the idea is compelling. For example, for those with a large farm, or anyone under labour pressure, the ability to clear a field rapidly and efficiently is important.

If that same farm wants to use hay or straw in small lots in many small and out-dated buildings, then the small conventional bale has many attractions. Bales for sale are more convenient as small bales

Plate 4.9 The Vicon bale has a high density and medium size. Here, four bales are being picked up at once for loading for long distance transport.

and handling efficiency may be achieved in a strapped package of 16 or 20 small bales. This is another system brought in (this time by McConnel) which has faded out.

Automatic trailer
There are now several different engineering approaches to the automatic trailer. The 'helter-skelter' arrangement from Germany, for example, seems to have died a natural death. The only one that has stood the test of time is the approach used by Sperry New Holland which is still built in half a dozen different versions but mainly for the American market. The basic idea is to free the baler from accumulators and allow the separated job of loading to be done by one man at a relatively high speed. This idea is simple enough; the result is a large, complex and expensive machine. Results, however, show that handling efficiency can be achieved and there is little doubt that, where sufficient scale exists, the costs can be justified.

Square bale
The large bale revolution was originally sparked off by Howard with its Bigbaler. It still has unique advantages. The baler is slightly more

Plate 4.10 Field handling of any bales is improved by handling several bales at once.

complex than some of the alternatives and the bale itself has a low density. The advantages of the machine may, however, be compelling.

Its workrate can achieve more than 20 tonnes an hour in the field. That is a genuine, quite common, figure which makes it faster than almost any other baling machine in the world. There are three other machines that will match it – two of which carry five-figure price tags. The bale itself will break up by hand into bundles and the bale can be dried with a blower.

Unfortunately, the bale is not a very dense package and it has almost faded out after a ten year history. The relatively dense Hesston bale has proved more attractive to people selling straw because its density lends to better transport efficiency.

Round bales

The round bale owes its success to two earlier machines; the old Allis-Chalmers round baler and Howard's square Bigbaler. But it survives on its own merits of the machine's simplicity and the bale's resistance to weathering.

The basic idea of the large round bale was a product of Mr Gary Vermeer's mind in trying to help farmer friends. The idea was to cut labour dramatically for the bale-handling job, but with a relatively simple and reliable machine. Vermeer released its 706 balers in 1972. The original Vermeer design was remarkably simple; it rolled up a bale like rolling up a carpet. Consequently, the machines do have a reasonable record of reliability in the field.

The round bale itself is also remarkable in its ability to withstand weather. It is, perhaps, important to put this into context. It is possible to suffer severe losses when leaving the large round bale outside to over-winter. When simple, common-sense rules are followed, however, this bale will stand well above 250 mm of rainfall with a low percentage loss.

The ability to withstand weather is such that the pressure is taken out of bale-carting. The bales can be collected when pressure of work allows rather than when imminent bad weather dictates. Incidentally, the 'rules' of storage state that hay 'thatches' better than straw.

Plate 4.11 Automatic bale sledges have never really caught on for big bales but they are a very cost effective and rapid way of clearing a field.

Plate 4.12 The great advantage of round bales is that they will weather comparatively well in the field.

Short straw is the worst risk. Bales are best put end to end in a long cylindrical line, preferably with the end to the prevailing wind and on a well-drained area with the air free to circulate round them.

Large bales
The largest of the original Vermeer balers brought into the UK were the 706s which were 7 ft in diameter and 6 ft wide. Most farm applications found such a bale daunting and the smaller sizes quickly caught on. The 605 size gives 6 ft in diameter and 5 ft in width. Sizes vary with manufacturer and deliver bales in the ranges of 5 × 4, 4 × 4 and 4 × 3. The larger the bale, the greater the efficiency in reducing the costs of man and tractor in the handling process.

Bale carrier
Table 4.4 examines an automatic bale carrier with deliberately conservative performance estimates. Even so, the machine appears attractive. The idea is to be able to drive out across the path taken by

the baler and pick up bales from the end. The forks at the front of the machine scoop up the bale and stand it on end.

A load of five bales is transported to the store and the bed chains are put in reverse to off-load. There is no hand work at all and the whole operation is on-the-move with pauses as distinct from stops. At unloading, the machine can leave bales in a line, touching end-to-end, which is the best way to store them outside. As the figures show, despite its lack of popularity so far, the big bale trailer can produce very low handling costs.

Conclusions

The major conclusion from the figures is that the 'bite' taken for handling must be a big one. It is for this reason that a process that picks up 20 or even 40 conventional bales at one run can compete with large bale systems. That is not really surprising as, on average, many large bales are equivalent to 15-25 conventional bales. It is worth noting what happens when several large bales are picked up at once (see the figures for the large bale mover that picks up five large bales at one run).

The conclusion reached is quite simple; for output a big machine-sized bite is required at each run. That means density and size count. This can be obtained by better loading of expensive tractors and there are a number of alternatives.

Taking the smallest bales first, there are several possibilities. Any system which can heap up bales into mini-stacks could give the advantage we are looking for. The only basic problem is that such an approach is bound to be complicated; it will involve complicated accumulators and at least some messing about on short runs for tractors.

Even the flat-eight systems can achieve this objective. Their biggest strength is that they are well-tried and thoroughly developed. Their biggest weakness is that a flat-eight of straw weighs 150 kg.

The figures in the table assume that the flat-eight is handled as such. When mini-stacks are built as, for example, with the Browns system, Farmhand's 40-bale trailer, Ritchie's bale carrier, or any other such carrier, the figures significantly improve.

Conventional bales from larger chambers obviously have output advantages, but the larger, often denser bale also has handling advantages. Systems which handle possibly 100 bales at one run – and there are several – can clearly cut down on tractor and man time and, therefore, costs.

This discussion must lead to considerable interest in the advent of the new, even-larger conventional balers from the USA. The new three-string bales can weigh nearly 50 kg. That may become another man-handling problem in itself at the end of the operation, but a machine-sized load is clearly going to give considerable potential.

These bales could be particularly important in the handling of straw to industrial plants. The one thing that comes out of experience is the importance of dry straw for processing. The moisture content of straw is the quality factor that worries processors more than all the other factors put together. We now have established really useful contractor operations working within the catchment areas of the processing plants. These are often based on the big, dense, Hesston bale. If the number of plants increases, however, their requirements will have an increasing effect on the way farmers supply straw.

Remember that they need dry straw for a long period. That means storage, on the farm. So most of that straw will be supplied by farmers, from their own storage, by their own transport in a bale that is dense enough to store under cover efficiently and transport efficiently. The 40-50 kg conventional bale may become more popular.

Plate 4.13 The genuine roll bale is very handy in livestock buildings.

The large bale has taken ten years to come of age. Current designs are much more dense and have more sophisticated controls and mechanisms. These machines now outsell conventional balers in the UK by about two to one.

DENSIFICATION

An argument often raised against further utilization of straw is its low density and the 'obvious' need to compress the material in order to obtain an economic load on transport. The idea is to cube pellet or wafer straw into a semi-flow, relatively dense product.

The problem is simple. A conventional bale is estimated to require (as diesel fuel to drive the tractor) about 2% of the energy available in the straw if it were burned for heat. With a wafering machine the figure is, in practice, often as high as 40%. That is not counting baling and handling to the press. The possibility of throwing away perhaps half of the value of the material in order to be able to use it might be alright with a high value material like uranium but with low value straw it is just not on. One further practical problem is that many of the imported wafering machines claim to have an energy consumption way below the figure quoted, one even claims the 2%, but, in practice, these claims have not been substantiated. Power consumption is often higher and output much less than claimed. However, research at the NIAE has shown that energy consumption of waferers can, technically, be brought down near to the figure for baling. Such a conclusion was reached following laboratory tests and commercial, high output machines have not yet been built. In theory it can be done. In practice, ask for evidence and a guarantee of performance.

From time to time, a bale densifier is produced that will compress a conventional or large bale to a fraction of its original length. A final bale density at least twice that of current bales is necessary to give maximum loading within size limits on lorries. These machines have a significant capital and running cost. There is also a separate operation involved and the end-user may not like the product.

It is inevitable that densification demands power use, even if some methods are better than others. It follows, therefore, that some expense will be involved and this can very easily undermine the economics of the operation the process was designed to help. So far, while very large amounts of money have been spent on machines to densify straw in some way, very little effect has been made on practical and economic progress.

ACTION PLAN — HARVESTING

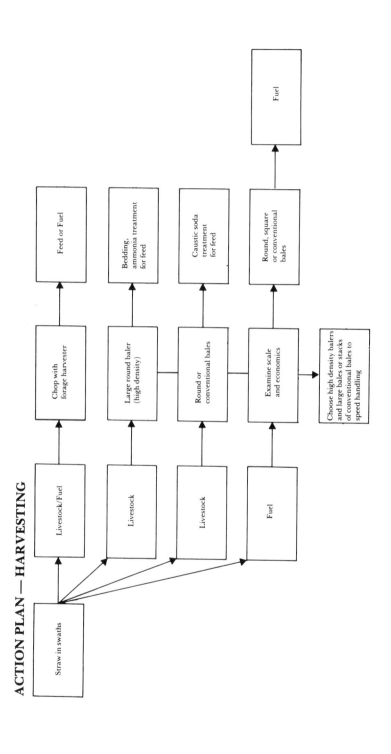

5
Straw feeding

THE POTENTIAL

It is as well to go back to square one to see what we are really aiming at when growing crops to feed cattle. Certainly, in a year when grass and other forage is short, straw is a necessary stand-by and any process which improves its digestibility is attractive. However, high mechanization costs cannot be borne by a system which only pays off in a bad forage year. To be attractive all the time the basic system has to be different. That fundamental difference occurs on a mixed farm.

Where a farm keeps stock which eat grass and where cereals are grown, there is a real potential to change. If you put in a few more cows per man, use more fungicide on the wheat, and more fertilizer, or squeeze the MOC in the dairy, these are all just nibbling at the problem. It will not radically boost income. Suppose, though, that the number of livestock stays the same; and that, instead of feeding grass, we feed treated straw. The arable area could be expanded. Farm running costs would be much the same because a hectare of cereals is not much more expensive to grow and harvest than the same area of conserved grass. The net result would be the same livestock production and similar costs but an increased cereal production. *That means the gross value of the extra cereal grain would be net profit.* That is a fundamental change and worth chasing. The question is how to treat the straw. We are not just interested in maintenance diets. We need production rations for rapidly growing beef animals and high yielding dairy cattle.

Plate 5.1 JF were the first in the field with a straw processor using liquid caustic soda.

STRAW AS A FOOD

Straw has been fed as a roughage ever since grain was first grown. In its untreated form it is a useful feed at low percentage inclusion in ruminant diets. Its physical characteristics, at low inclusion rates, help rumen function. Chemically, however, the food is of relatively low value in its untreated form.

Table 5.1 gives a brief summary of a few crops compared with straw on the basis of dry matter, metabolizable energy (ME) and crude protein. Clearly, on the face of it, cereal straws are not capable of sustaining production diets at high levels of inclusion. The animal could just not eat enough and digest enough in a day. This statement covers up two significant functions of digestion which are important in the discussion of the treatment of straw. The digestibility of straw, however it is measured, and the throughput (or dilution rate in the rumen) will determine how much straw can be usefully eaten. Dry

Table 5.1 Dry matter, energy and protein supply for a range of crops commonly grown in the UK

	Dry matter (g/kg)	Metabolizable energy (MJ/kg dm)	Crude protein (g/kg)
Wheat	860	14.0	124
Straw	860	5.7	24
Barley	860	13.7	108
Straw	860	7.3	38
Oats	860	11.5	109
Straw	860	6.7	34
Potatoes	210	12.5	90
Sugar beet	230	13.7	48
Fodder beet	130	12.6	77
Kale	160	11.1	137
Maize	190	8.8	89
Grass	200	12.8	265

Matter Intake and Digestibility are key factors in the design of all diets but especially so with straw because of its high bulk.

In fact, straw varies considerably in its value as a feed, both physically and chemically. The cereals vary significantly between species and also between varieties.

Table 5.2 The nutritive value of straws of different species

Species	ME (MJ/kg)	DCP (g/kg)
Barley	5.8 – 7.3	9
Oats	6.7 – 6.8	10
Wheat	5.6 – 5.7	1

Source: AP No. 11 (MAFF).

As Table 5.2 shows, oats and barley have both higher energy and protein than does wheat straw in its untreated state. Varieties also show differences but not so marked.

Table 5.3 The variation of energy content within a species (S barley)

Variety	Est. ME (MJ/kg)
Mazurka	7.5
Maris Mink	7.2
Universe	7.1
Proctor	6.9
Julia	6.7
Armelle	5.9

Source: Palmer (1976a).

It is also known that the part of the plant also affects nutritive value. The light components at the top of the plant are much more useful than the stem and lower parts. So, the cutting height of the combine and the length of stubble not only affects the quantity of baleable straw but also its quality (Table 5.4).

Table 5.4 The variation of nutritive value of barley straw according to height of cuttiing.

Portion	Approximate analysis (g/kg)		
	CP	CF	Ash
Top third	46	369	63
Middle third	30	480	43
Bottom third	29	502	43
Mean	35	450	50

Source: Wilson (1975).

Plate 5.2 The inside of the JF processor mixer chamber.

A paper prepared by the ADAS Nutritionist F.G. Palmer also outlined a number of other factors that affect food value.

Time of sowing

It is generally accepted that winter sown cereal straws have a lower food value than spring sown ones. For barley the energy values given in Advisory Paper No. 11 (MAFF) are 5.8 MJ and 7.3 MJ ME/kg respectively, which reflects the higher fibre content in the winter sown crops. Later data (Palmer 1976) showed that there was little difference in fibre content between the two and both had a ME content of about 6.0 MJ/kg. Generally, the farmers' view that spring sown straws are likely to be of greater food value must be right in most circumstances.

Plate 5.3 Applying Granstock, which is basically urea with added salts and vitamins, with a watering can to a straw bale.

Climatic conditions

Straw produced in 1976 and 1984 appeared to have a higher nutritive content than normal as a result of the drought conditions, although the data available are limited. It is reasonable to assume that nutritive value will vary from region to region because of the different climatic conditions.

Physical properties

It is known that some growth regulators work by shortening the stem and increasing the diameter. At a given combine cutting height this probably does increase, by proportion of the upper more valuable parts, the feed value of the cut material if the light parts are retained.

Work at Nottingham University in the late 1960s and early 1970s showed that milled straw had about 10% more available energy than chopped straw. Fine chopping of straw also pushes up dry matter intake. Palatability drops as dustiness or dryness increases (DMI). Straw as a meal is more likely to affect DMI than is straw in pelleted form.

CHEMICAL TREATMENT

While the basic constituents of straw are well known, the exact molecular structure of these molecules is very complex and has never been precisely described. For that reason, we do not know precisely what happens when alkalis are used to 'treat' straw.

Ruminant animals are capable of digesting cellulose and hemicelluloses, which are cell wall constituents (Tables 5.5 and 5.6), by the

Table 5.5 Composition and nutritive value of cereal straws (g/kg dry matter)

Crude protein	Cell walls	Cellulose	Hemicellulose	Lignin	Ash
38	840	440	330	80	53

Table 5.6 Typical cellulose, hemicellulose and lignin content of grass and straw

Forage	Cellulose (%DM)	Hemicellulose (%DM)	Lignin (%DM)
Fresh grass	25	28	4
Straw	41	29	11

action of rumen bacteria. In highly lignified materials (e.g. straw), these potentially digestible carbohydrates are protected by their association with indigestible lignin. This results in a food of low digestibility, and low intake potential which can supply only a small part of a productive animal's energy requirement. Chemical treatment with alkalis, such as sodium hydroxide, can modify the lignin – carbohydrate complex in straws making the cellulose and hemicellulose fractions accessible to bacterial digestion and thus increasing their digestibility and energy value.

Effect of sodim hydroxide on the composition of straw

The effect of alkali on cell wall structures is probably the result of hydrolysis of hemicellulose linkages, allowing a physical swelling of the encrusting lignin framework and thereby exposing the digestible components to bacterial attack. Sodium hydroxide is an inorganic compound and therefore its addition to straw results in a reduction in crude protein, organic matter and gross energy, by a simple dilution effect.

In fact alkali treatment has a greater effect on highly lignified, low digestibility material than on comparatively 'good' higher digestibility straws. The effect of treatment then, will be to narrow the gap between the digestibilities of different materials. As wheat straw chops more easily than barley straw, it is sometimes used in preference to barley straw when treatment will result in very similar products. After treatment, the barley straw might have a marginal extra value (Table 5.7).

ALTERNATIVE TREATMENTS

There are potentially several different basic treatments for straw. Alkalis soften the lignin structure and these are the chemicals used in practice at present. Of these, sodium hydroxide (caustic soda) is the most effective. Ammonium hydroxide (ammonia dissolved in water) is a close second. Calcium hydroxide can be used but is very slow and relatively weak in its effect.

Acid hydrolysis could be more effective but has not yet seen development at the farm level. Other chemicals may have effects by encouraging rumen bacteria. Given time and reasonable temperatures, urea is converted by bacteria on straw into ammonia which then acts to process the straw as an alkali. When eaten immediately

Table 5.7 Nutritive values of straw treated by various processes

(a) Industrial process. DOMD value (in vitro). Sodium hydroxide at 4% rate

	Untreated	*Treated*
Wheat	49	66
Barley	50	66
Oats	52	70
Rye	44	68

Data: F. Rexen *et al.*, Denmark.

(b) Laboratory process. Digestibility in rumen. Sodium hydroxide at 3.3%.

	Untreated	*Treated*
Maize cobs	46	60
Wheat	51	67
Paddy rice	57	69
Soughum Stova	58	68
Sugar cane tops	56	65

Data: Chandra *et al.*, India.

(c) Laboratory process. Enzyme digestibility. Sodium hydroxide at 5% rate at room temp. for 30 days

	Untreated	*Treated*
Alfalfa (lucerne)	53	62
Barley	37	73
Bean	52	65
Fescue	37	62
Oat	33	63
Ryegrass	40	65
Rice	29	62
Wheat	37	62

Data: Vice. USA.

after treatment urea treated straw has a better nitrogen/energy balance which will encourage rumen bacteria. ICI's Granstock is based on a very high concentration of urea in solution plus added vitamins and minerals. Strawmix is a Scandinavian development based on a high inclusion rate of molasses to encourage a high level of bacterial activity.

Strawmix

Method
The addition of molasses to chopped straw is followed by inclusion in a specific previously formulated complete ration. Straw chopped to 5–7 cm.

Plate 5.4 Weighing out solid caustic soda for putting into a mixer wagon for 'instant' treating of straw.

Chemistry
A chemical reaction on the straw is not involved. The process involves setting up a relatively constant condition in the rumen with the molasses present to stimulate a rapid bacterial attack on the straw.

Likely ME
ME of straw mixed with molasses will come out in the region of 7 MJ/kg. Kongskilde claim that Strawmix including the rest of the complete ration would have an ME of at least 7.5 MJ/kg on a fresh weight basis or over 10 on a dry matter basis.

Capital cost
Complete plant would cost upwards of £20 000 depending on capacity.

Chemical cost
Molasses is applied at 25% of the total mix and 45% of the straw. At £60/tonne for molasses this would cost £27/tonne of straw treated. To this, the rest of the cost of the ration would have to be added.

Plate 5.5 Applying liquid caustic soda to straw being chopped in a tub grinder as a contractor's service.

Labour
There are no special chemical dangers to staff involved.

Time
There is no special chemical reaction involved so there is no time delay. Labour is involved handling in and out of the Strawmix plant.

The Strawmix process was really a philosophy based on the idea that a source of highly available energy would lift rumen bacterial activity to a point where useful extra levels of use of cheap fibrous foods could be exploited. The diet produced is really a complete diet intended to be a complete mix which is relatively constant while allowing for marginal adjustments on a day to day basis.

Urea – Granstock

Method
Granstock, which is a solution containing urea, minerals and vitamins, is added to straw. A uniform addition is not necessary, application with a watering can on to bales is adequate. Straw: normally long, chopping is not necessary.

Chemistry
A chemical reaction on the straw is not involved. The urea is a source of NPN which encourages rumen bacteria.

Likely ME
Will get about 10% increase in energy but does give increase NPN. May go up by up to one unit ME.

Capital cost
A watering can.

Chemical cost
About £95 per drum of 210 litres. Application is normally at 30 litres/tonne, i.e. about £14/tonne treated.

Labour
Handling the straw bales and watering on the liquid supplement. The product is very safe to handle.

Time
No time delay is necessary.
   Granstock can be applied to straw immediately before feeding. Application can be via a watering can or sprayer. Even application is not highly critical. Granstock relies on the addition of urea as a nitrogen source for the rumen bacteria. It also adds minerals (such as sulphur and phosphorus) and vitamins which are known to be required in high straw diets.

Urea – as a processor

Method
Urea applied to straw will change to ammonia and act as an alkali. At least 22°C is needed for two weeks. Straw: normally long. Chopping will reduce gas movement.

Chemistry

Urea is changed to ammonia by bacterial action. The ammonia dissolves in the water in the straw to produce ammonium hydroxide which softens the lignin structure.

Likely ME

If temperatures are high enough, the result will be comparable to ammonia treatment with an ME of around 7.

Capital cost

Low. Sealed storage can be arranged with polythene.

Chemical cost

Urea costs about £15 per 50 kg. Inclusion at 5% would cost £15 per tonne of straw treated. At 2% it would cost £6.

Labour

Farm labour is involved and there are no special safety precautions for labour.

Time

At least two weeks at 22°C.

Caution

If application is above 2% and temperatures are not above 22°C for two weeks, conversion of urea to ammonia does not take place, and there are toxicity problems.

Generally speaking this method of treatment is not successful under UK climatic conditions. However, at the 2% rate, urea is a useful addition of NPN to straw. Beef cattle can take the whole of their protein requirement in this way.

Ammonia – ammonium hydroxide

There are basically three methods of working with ammonia. Ammonia can be injected in anhydrous or aqueous form into a stack or it can be circulated in anhydrous form in an oven.

Ammonia is a very corrosive substance with some wicked properties but very small traces in the air (three parts/million), can be easily smelled and so there is good safe warning. At higher levels though, say 20 ppm, it will do damage to human lungs.

Plate 5.6 The basis of the 'Strawmix' process is a chopper plus the application of molasses.

Aqueous ammonia

Method
Aqueous ammonia. Injection of stack of bales or chopped straw.

Chemistry
Ammonium hydroxide (i.e. aqueous ammonia) is an alkali which chemically attacks straw.

Likely ME
Six to eight.

Capital cost
Very low. Plastic sheet to seal the stack. Net to prevent wind damage. Possibly £10/tonne or less if re-use of plastic sheet is possible.

Chemical cost
About £20/tonne treated at 3%.

Labour
Labour is involved in getting the straw into and out of the stack. Ammonia is normally handled by contractor's staff.

Time
Time delay for the process to work would be 2-4 weeks in warm weather. It is not recommended in cold weather.

Aqueous ammonia has attractions because it does not need pressure vessel transport and storage. However, the Norwegians have not had much success with the method. They have had problems with inconsistent results; some have been quite as good as those with anhydrous ammonia, others have had little or no effect.

It has been difficult to achieve uniformity in the stack and there have been effluent problems. However, there has been quite a bit of development work done by a UK company which claims different and more successful results. The convenience factors may be compelling under some circumstances.

Anhydrous ammonia

Method
Ammonia-anhydrous. Injection of stack (see Fig. 5.1).

Chemistry
Ammonia gas dissolves in the water in the straw to produce ammonium hydroxide which is an alkali and chemically attacks the straw.

Likely ME
Generally in the region of 7–7.5.

Capital Cost
Very low. Plastic sheet to put under and over the stack. Net to prevent wind damage. Possibly £4/tonne or less if re-use of plastics is possible.

Chemical cost
£17/tonne of straw to apply at 3.5% which is probably necessary. NB: Norwegians apply 4%.

Labour
Farm labour is involved with building the stack and handling the

plastic sheet. The ammonia would normally be handled by contractor's staff. Care should be taken when opening the stack and two people should be present.

Fig. 5.1 Principles of the 'stack method' for ammonia treatment of straw (Norsk Hydro/NOFO, 1977).

Time

In cool weather allow 6–8 weeks for treatment. In hot weather 2–4 weeks will be sufficient.

Anhydrous ammonia has to be stored in pressure vessels and expansion valves get very cold, so much so that flesh may stick to them. The aqueous form is a corrosive alkali. Having said all this, ammonia is generally safer than caustic.

All ammonia methods have the special advantages of adding NPN to the straw. With good results, ammonia straw should be equal on both energy and total nitrogen to reasonable quality hay. A further advantage of ammonia treatment is that there are none of the side effects associated with caustic additives.

Stack treatment with anhydrous ammonia was pioneered by the Norwegians with the help of their local suppliers, Norsk Hydro (who have taken over Fisons Fertilisers in the UK). It does work reasonably well with low capital costs. However, there is quite a bit of labour involved in stacking and unstacking and the polythene cover must be kept intact against wind and vermin.

Results in the UK have been more variable than the oven method. Despite its disadvantages it is a useful low-cost method with a low capital commitment. Many who are interested in processed straw try this method to see how they get on with the idea.

Anhydrous ammonia

Method

Ammonia-anhydrous (gas). Oven treatment.

Chemistry

Ammonia gas dissolves in the water in the straw to produce ammonium hydroxide which is an alkali and chemically attacks the straw.

Likely ME

About 30% better than stack treatment. Generally in the region of 8–9. Results are more consistent than with stack treatment.

Capital cost

Depreciation will cost £5–£8/tonne. Capital cost will be £6000 up to perhaps £12 000.

Plate 5.7 Applying caustic soda in the solid form with a suitably designed forage harvester with a separate blower.

Chemical cost
Three per cent ammonia costs about £15/tonne of straw. Electricity at £5/tonne.

Labour
Ten min/day with latest ovens with large round bales.

Time
The treatment cycle is generally about 22h.

The oven method was pioneered by the Danes and Danish machines are available in the UK. However, significant progress has been made in the design and automation by the British firm, Straw Feed Services, who are now turning the tables with exports to several countries.

The main advantage of the method is uniformity. A consistently high-value material is produced at reasonable cost. With big bales or pallet loading of conventional bales, labour can be kept reasonably low.

Caustic soda – sodium hydroxide

As a general guide ammonia treatment is most commonly used for long straw and caustic for chopped. However, both aqueous and anhydrous ammonia have been used for injecting chopped straw but the general conclusion at present is that with such treatment it is difficult to get even distribution throughout the heap.

It is also possible to tub-grind big bales treated with anhydrous ammonia. However, output might be lower than with untreated straw. All common methods of caustic soda application work with chopped straw.

Caustic soda (sodium hydroxide) is the strongest of the alkalis used for processing straw and it is marginally the best of those in common use. It can produce better results than ammonia. It can be applied as liquid or as solid.

Liquid caustic is very dangerous and will burn flesh immediately on contact. In solid form it burns only after going into solution in sweat or in the moisture in the eye. Solid is, therefore, much safer. Liquid processes can be carried out with ordinary farm labour provided codes of practice are strictly adhered to.

One of the drawbacks with caustic is that the sodium has to be expelled from the animal and so it tends to drink and urinate more. This can be avoided by feeding salt-free minerals. (Most mineral supplements contain a high proportion of common salt – sodium chloride.)

Liquid caustic soda

Method
Normally chopped straw is used with continuous process machinery, i.e. the caustic is applied at the same time as chopping in a tub grinder, chopper or forage harvester.

Chemistry
The application of liquid caustic soda to straw causes attack on the lignin and hemicellulose and will soften them so that bacterial action in the rumen will increase ME substantially.

Likely ME
From 8 to 9.

Capital cost
The JF machine is now only built to special order but would cost several thousand pounds new or used. A tub grinder would cost £6000–£60 000. These costs are often operated by contractors. Application at a forage harvester would cost a few hundred pounds to convert the forager.

Chemical cost
Caustic soda is normally applied at 5% by dry weight. A 32% solution will probably cost about £40 for a 225 litre drum. This would be about £25/tonne treated. The drums have a high deposit on them. However, buying in bulk can cut this price significantly, giving a cost of under £10/tonne treated, but a bulk tank has to comply with H and S standards.

Labour
Involved in handling the straw and running the machinery. With care and discipline, the process can be carried out quite safely with farm labour. Contractors are available in some parts of the country.

Plate 5.8 Using a forage harvester to apply solid caustic soda to whole crop barley.

Time
A time delay of 3 days is necessary for the reaction to be completed.
The above summarizes the overall situation. JF are doing quite well in the West Country particularly with contractors processing straw for dairy cows for whom sodium hydroxide has something of the same effect as bicarb as well as processing the straw. Salt-free minerals should be fed so as to reduce urine production.

Generally speaking, applying liquid caustic when cutting with a forage harvester is regarded as potentially dangerous. With tub grinders it has been difficult to get the application of caustic soda even enough for satisfactory results.

Solid caustic soda

Method
Granular caustic soda is added to straw at 5% by dry weight and mixed thoroughly. The reaction is very rapid and virtually complete within a few minutes.

Chemistry
The reaction is exactly the same as with liquid caustic soda except that when solid dissolves in water it releases a great deal of heat. This heat helps the reaction and speeds it up.

Likely ME
Eight to nine.

Capital cost
Depends on the system. Solid can be applied on a forage harvester when the applicator will cost about £800. Solid can also be added in weighed amounts into a feed mixer wagon.

Chemical cost
Solid caustic will cost about £200/tonne. This works out at about £10/tonne treated. Bought in bulk the price could be £187/tonne.

Labour
The straw has to be moved but the chemical itself involves little work because it is concentrated and there is no water to carry out. Goggles and gloves should be worn when handling the sacks of chemical.

Time

A time delay after thorough mixing is not necessary (provided there are no lumps in the caustic soda).

The text above looks at solid caustic soda; ICI's brand is known as Pearl. Unlike flake caustic, Pearl granular caustic is almost dust free and much easier and safer to handle. Any auger-type machine can be used to apply the granules safely and evenly. The auger works the granules over chopped straw into which it progressively dissolves.

The application can be to batches in a feed mixer wagon, which has proved very successful, or on forage harvesters which have a separate chopper and blower and an auger between the two.

Applications on forage harvesters have been carried out with a Berwyn Meterite applicator at 5% per dry tonne and results are good.

Cocktails

Method

Granular caustic is added to straw in a feed mixer wagon. Molasses, urea, minerals and vitamins are also added.

Chemistry

All the reactions involved with the substances used separately apply to the cocktail.

Likely ME

Nine to 10.5 or higher.

Capital cost

None if the feed mixer trailer is already owned.

Chemical cost

Caustic soda at 5% costs about £15/tonne treated. Addition of urea as urea would cost £15 and molasses about £30/tonne treated.

Labour

Handling the materials into the mixer trailer. Goggles and gloves to be worn when handling the granular caustic.

Time

A time delay is not necessary after thorough mixing.

The idea behind 'cocktails' is not to enter the argument about which process is best but to apply the lot to one sample of straw. The cocktail could be anything, but trials carried out in 1981 by the author started with caustic soda (the most active alkali) in granular form, molasses and urea and minerals and vitamins.

The aim was to get the maximum softening up of the straw and to assist the rumen bacteria further by adding molasses as a readily available energy source and urea as a readily available NPN source.

The process was carried out in a feed mixer wagon and feeding results were good. A trial batch of Hereford cross steers gained 1.7 kg per day on 50% cocktail straw plus sugar beet pulp and silage but no grain. Further feeding of commercial sized groups of several hundred cattle has consistently achieved figures of 1.3–1.5 kg per day.

General comments

One of the difficulties in comparing the different methods is that the in vitro digestibility test gives an unreliable ME figure. Generally, (but not consistently), it under-estimates the real *in vivo* (through the animal) figure. However it is the only figure that is widely available and, as a result, the estimates provided in the tables should be treated with caution.

It is true that increases in dry matter intake are often experienced when processed straw is fed. If cattle are switched to a high level of straw from a diet with less (or none) then they must have the capacity for greater intake if they are to perform as well. For instance, stores normally kept on high roughage diets can often do very well and fatten at high rates when put on to processed straw or 'cocktail' treated straw. Any real comparison should be based on how the livestock perform and there is quite a bit to learn on how best to incorporate processed straw of any type into a ration.

Livestock performance will also vary with the ration that the straw is fed with; some rations work better than others. For example, alkali-treated straw works particularly well with high concentrate diets fed to dairy cattle.

It is important to note that all high-level straw diets need special mineral and vitamin additions. Sulphur and phosphorus are especially important and so are magnesium and copper. Selenium and vitamin E should also be incorporated at higher rates.

Stock fed on caustic-treated straw should be fed salt-free minerals to reduce the input of sodium.

Code of practice for handling solid caustic soda in farm use

Storage
- Solid caustic soda should be stored in sealed sacks or containers held in a confined store not accessible to children or other 'unauthorized' personnel. Handling should, wherever possible, be in bags not heavier than 25 kg.

Handling
- It must be remembered that even small particles of caustic soda will burn the skin. If washed immediately, such damage is unlikely to be serious. Contact with the eyes, however, is potentially serious and all operatives must be fully aware of such danger.
- Goggles plus PVC or rubber gloves *must* be worn at all times when handling the sacks.
- If a sack is to be opened during handling, gloves, goggles (not visor) and suitable protective clothing should be worn. 'Suitable' means that granules will not be able to penetrate the outer layer nor get caught in gloves, wellingtons, etc.
- When handling machinery or products which contain caustic deposits or through which caustic soda has passed, protective clothing should be worn – that necessary will be defined by the risk but the normal minimum will be gloves and goggles.
- Unnecessary personnel should be kept well away from the region of work.
- Adequate quantities of clean water should be available in the store and wherever the material is to be handled or used. Eye wash bottles or boric saline solution must be available. Any person who has any contact of caustic soda in the eye should be checked over by qualified medical staff as soon as possible, but this must not delay immediate irrigation of the eye, preferably with boric saline solution.
- All tractor drivers involved in the operation at any point will be supplied with gloves and goggles and with their own container of at least 4.5 litres (or 1 gallon) of clear water.
- Although the danger of contact with straw treated with caustic soda is not high, access to that straw by unauthorized personnel should be restricted for at least 12 h after the treatment took place.

COCKTAILS IN DETAIL

Methods in practice

Some early trials involving forage harvesters used liquid caustic soda. The problem with liquids is a logistical one; trouping a tanker round the fields has its problems. Liquid caustic is dangerous to handle. Straw treated with liquid with a *total* moisture content of over 30% has a fire risk if left in heaps of over 10 tonnes. Liquid treatment also involves a delay of 3 days before the reaction is complete.

Trials with solid or 'Pearl' caustic appear to offer a solution. In practice, the reaction does go well with only a short delay. The problem is that with most forage harvesters it does lead to a safety problem. From a materials handling point of view, to pick up the straw, chop it, treat it and put it into a transport trailer in one operation has its attractions. The problem is that a limited quantity of small particles of solid caustic do get out into the air and that thin cloud of caustic dust is, without doubt, potentially dangerous.

Field research has centred on either controlling that cloud or processing the straw in the farmyard. This can, according to the latest investigations, be eliminated provided a suitable forager is used. A

Plate 5.9 Treating straw with a cocktail based on solid caustic soda to produce a complete ration using a contractor's service.

forager with separate blower and an auger between the cutter head and fan appears to solve the problem. However, the only forager of this type currently on the market is from Bamfords of Uttoxeter. This process opens the way to applying both solid caustic (as granules *not* flake) and solid urea at one pass with a forage harvester. It has been done with good results right through to feeding on both straw and whole crop barley.

An alternative is to handle and store straw conventionally and then chop it immediately before use. It can then be treated 'instantly' with solid caustic soda applied in a feed mixer trailer. The speed of the reaction depends on the emission of large quantities of heat of dissolution by the caustic soda as it dissolves in the water in the straw. Some mixer trailers are better than others at mixing and can achieve an adequate reaction in 2-3 min. Others may take up to 10 min. The test is a dramatic change of colour to a golden yellow and a reduction in volume of up to 50%.

Addition of the rest of the ration, plus urea and molasses as in the brief description and text above gives the completed ration which can be fed immediately to dairy or beef cattle.

Procedure for cocktail treatment of straw

- Most or all of the straw for one mix should be loaded first into the mixer wagon. Chopped or milled straw should be at any convenient length such as 3–7 cm (say 1–3 in) range. The mixer should be running. The moisture content should be brought up to 20–25%.
- The caustic soda granules should be added next and all particles must pass a 6 mm (0.25 in) sieve. After a short mix of as little as 3–10 min, the rest of the cocktail components can be added. The caustic mix time is judged by (a) a dramatic colour change to golden yellow and (b) a reduction in volume of 30–50%.
- The molasses and urea can now be added and the usual mix time allowed before feeding. A long delay is not necessary; the evidence is that provided there were no hard lumps in the caustic soda, there will be no unabsorbed caustic that could harm the stock.
- The caustic soda, molasses and urea can be added via a purpose built dispenser, a controlled tip loader bucket or added by hand via a bucket or scoop. Normal codes of practice should be followed.
- The rest of the diet such as silage can now be added.

THE DESIGN OF DIETS

The purpose of straw feeding is not necessarily to maximize production (although high levels of growth and yield remain fundamentally necessary). There are two real targets of which the first is obvious enough although it is not the most important.

Straw feeding offers the scope to feed cheaper diets. Field experience with commercial beef has shown liveweight gains in the 1.0–1.5kg per day range with cost savings of up to 40% (see below).

The second potential is to re-balance the economy of the farm. For a mixed farm with livestock based on grass production and cereals with inadequately used straw, the cost of growing and harvesting cereals or grass is very similar. Cereals will be a little more expensive but not dramatically so on a per hectare basis. If grass is switched to cereal production, total costs on the farm will be roughly the same, maybe a little higher. However, the gross output of the farm will jump by the gross value of the extra grain as the level of stocking can be maintained with stock fed on a straw based diet supplemented by some grass product. In this situation, the *net profit* of the farm goes up by the *gross value* of the extra grain. An alternative is to hold the cereal production and carry an increased stocking level. Again, net profit will be approximately equal to the gross value of the increased production.

By any standards that is a compelling argument; to increase the profitable production of the farm as a whole. It is this argument which will influence what sort of ration is fed to the stock and what the ultimate production objectives will be. The actual target in practice is likely to be a high level (but not the highest) of production with a cheap diet based on a high proportion of straw so as to release hectares for increased cereal production or to carry increased stocking level. Waste can be eliminated and production can rise. Profit potential depends on production at reasonable cost. With straw feeding it is possible to have both.

Measures of feed value

Generally speaking, ME has not been a highly accurate means of predicting the food value of roughages and it has sometimes been very misleading with respect to treated straws. Even the *in vitro* (in the laboratory) testing for digestibility produced results which varied from laboratory to laboratory. A much better test is the *in vivo* (in the animal) test to give DOMD value.* G.I. Givens of the ADAS Nutrition Chemistry Feed Evaluation Unit, Stratford, gave this summary at a conference at the RASE, 1st November 1984.

'Whilst the *in vitro* test provides a reasonably quick and relatively cheap means of evaluating straw, the application of the results in practice must assume that there is a good relationship between the *in vitro* DOMD value and the corresponding *in vivo* value. However, work at the Feed Evaluation Unit has recently statistically examined the relationship between *in vitro* DOMD and *in vivo* DOMD values for straw examined there. The data show that there is not a statistically significant relationship between them and whilst in general the *in vitro* values tend to be higher than the *in vivo* values, the bias is not consistent. The ability of the *in vitro* method to predict the *in vivo* DOMD straw treated with sodium hydroxide appears to be even worse than with untreated straw. The reasons for the lack of a consistent relationship between the *in vitro* and *in vivo* measurements are not fully understood, but the one important factor is likely to be related to the sample preparation prior to the *in vitro* test. The standard method of preparation is to hammer mill the straw to pass through a 1 mm screen. It is likely that this very vigorous processing presents the microbes in the rumen fluid with a material more readily digested than straw in the form normally fed. Another problem, particularly with alkali treated straw, is that during the treatment process some of the lignin is solubilized but remains indigestible by the animal. However, because of the analytical procedures in some *in vitro* methods this material is recorded as having been digested. Despite the lack of good agreement between the *in vitro* and *in vivo* results, animal performance data predicted from *in vitro* DOMD measurements have been shown (Adamson and Bastiman, 1984) to agree reasonably well with actual performance. This may be partly due to the *in vitro* DOMD reflecting the voluntary intake characteristics of the straw. To date no statistically significant relationship has been found between any of the laboratory techniques and the *in vivo* measurements (see Barber and others 1984a) and it must be concluded that at the present time there is no way of predicting the *in vivo* DOMD or ME value of straw by laboratory measurement. Some newer sophisticated laboratory techniques are beginning to emerge such as the cellulase digestibility of isolated straw cell walls

---

* Digestible Organic Matter in the Dry Matter.

and the use of near infra red reflectance analysis. The latter technique, in particular, has the ability to examine samples very rapidly for a wide range of parameters and is showing promise although the development of the method is still in its very early stages. This would allow a much more comprehensive characterisation of straws than hitherto on a routine basis.'

Conclusion

At present ME is the easiest, fastest and cheapest measure we have got. So we use it. Remember, however, not to treat it as being too accurate. Try a diet, monitor performance and make empirical (practical) adjustments as you go along.

Caustic or ammonia treatment

Generally, dairy cows in the UK have enough non-protein nitrogen (NPN) in their diets. Under some circumstances, more NPN may be a disadvantage. On the other hand caustic treated straw contains sodium. Sodium has been shown (as sodium bicarbonate, for example) to increase dry matter intake and improve butter fat. Therefore, many observers take the view that caustic treatment of straw has advantages for dairy cows.

Ammonia treatment, however, not only acts as an alkali to 'soften' the lignin but also supplies extra NPN which can be used by ruminants, especially beef cattle, growing heifers and dry cows.

Generally, caustic treatment uses chopped straw which produces a feed easier to ration for the controlled diet of dairy cows. Ammonia treatment normally at least starts with long straw which is conveniently fed to beef, heifers and dry cows. Ammonia straw can, however, be chopped after treatment.

There are, then, some natural pointers as to which to use. However, in practice there is a good deal of overlap. Caustic treated straw is widely used for beef, and ammonia treated straw for dairy cows. Generally, the caustic treatment is more suitable for highly productive beef or dairy, with ammonia treatment more suitable and economic for less productive stock.

CAUSTIC TREATED STRAW – PRACTICAL DIETS

Gordon Newman, speaking at the RASE Conference in November 1984 gave the following practical examples of the use of caustic treated (sodium hydroxide) straw. Mr Newman sees caustic treated straw as having four main functions in dairy cow diets:

(a) As a method of saving silage;
(b) as a buffer feed at grass;
(c) as a physical conditioner;
(d) as a chemical buffer to assist rumen fermentation and increase appetite.

Proven practical examples of each function are discussed below.

(a) Saving silage

Untreated barley straw should be fed in large quantity to dry cows at autumn and late summer grass. After dry cows are housed, the straw should be supplemented by urea, molasses and magnesium (Table 5.8). Alternatively ammonia treated straw plus magnesium in water or caustic treated straw with grass silage plus magnesium are highly successful. With its low calcium and nitrogen content, prepartum feeding of straw prevents metabolic disease later.

Cows in late lactation, such as spring calvers in the autumn or September calvers in the spring can tolerate replacement of most of their silage by caustic treated straw.

Treated straw can be self-fed to lower yielders. Acidotic high yielders receiving large quantities of 'concentrates' may prefer alkaline straw to acid silage and yields are then depressed by reduced energy intake. Caustic treated straw is excellent underneath wet grass silage – not only is it absorbent but also nutritionally similar to the silage after soaking in effluent.

(b) As a buffer feed at grass

Whilst butterfat is a major constituent of milk price, provision of adequate digestible fibre in the diet is essential. Treated straw is an excellent method of maintaining butterfat in summer and increasing stocking rate so that more first cut silage can be made. Several herds on this diet have not fallen below 4% BF all season.

Table 5.8 Practical diets for different stages of lactation

	Dry cow diet	
	Feed kg	DM kg
Caustic treated straw	ad lib, say	8.0
Grass silage	10.0	2.0
Molasses	0.5	0.3
Calcined magnesite	0.06	0.06

Table 5.9

	In calf cow (15 litres)	
	Feed kg	DM kg
Grass silage	15.0	3.0
Caustic treated straw	ad lib, say	6.0
Molasses	4.0	3.0
M Gluten feed	3.0	2.7
White fish meal	0.3	0.3
Urea	0.1	0.1
Mineral	0.1	0.1
Total		15.2

Table 5.10

	20 litre Cow on grass by day	
At night	Feed kg	DM kg
Caustic treated straw	ad lib, say	6.0
Molasses	1.0	0.7
Calcined magnesite	0.1	0.1
Total		6.8

It is essential that treated straw fed with grass should be made palatable with molasses and that feeding is 'compulsory' – it is no use allowing cows to walk past a trough full of straw on their way to pasture and expecting them to relish it!

(c) As a physical 'conditioner'

When diets with a high physical density are fed (maize silage is a good example), displaced abomasum is a common problem. This is eliminated by the inclusion of treated straw.

Table 5.11

	Heifers yielding 21 litres	
	Feed kg	DM kg
Maize silage	ad lib	10.2
Caustic treated straw	4.0	3.4
Molasses	1.0	0.75
M Gluten feed	2.0	1.8
Soya 44%	1.0	0.9
W fish meal	0.75	0.7
Urea	0.15	0.15
Mineral	0.20	0.2
Total		18.1

This diet would not be possible without straw which seems to play a 'lucerne hay' role in physical terms.

(d) As a chemical 'buffer' to increase appetite

Maintenace of appetite in freshly calved cows is essential in order to reduce concentrate costs as well as improved milk quality and fertility. Caustic treated straw, with its low substitution rate and low nitrogen content can be used either as a constituent of dairy cake or put directly into the ration. It is more effective than sodium bicarbonate.

Table 5.12

	Fresh calvers yielding 35 litres	
	Feed kg	DM kg
Grass silage (ll ME)	ad lib, say	9.0
Caustic treated straw	4.0	3.3
Molasses	2.0	1.5
Wheat	3.0	2.5
M G Feed	3.0	2.7
W fish meal	1.0	0.9
60% M G Meal	1.0	0.9
Mineral	0.2	0.2
Total		21.0

The high dry matter is achieved if the cows have been 'steamed up' (or down?) on the dry cow ration in Table 5.8.

Mr Newman's summary is that the strategic use of untreated barley straw and caustic treated wheat straw in dairy cow rations is allowing dairy farmers, whose total production is restricted by quota, to increase stocking rates and reduce feed costs. The effects on herd health and milk quality are useful bonuses. Although straw treatment is currently carried out by a chain of independent contractors with JF processors for Mole Valley from Sussex to Cornwall, they are restricted to conventional bale size and we all look forward to a new generation of contractors' machines which can treat large Hesston bales with equal precision.

AMMONIA TREATED STRAW – PRACTICAL DIETS

Mr B. Bastiman of the High Mowthorpe EHF presented a paper at the RASE conference on straw in November 1984. The results are summarized below.

For many years EHFs have been involved in work on the inclusion of straw in beef rations. In the early years this covered inclusion of chopped, ground or industrially nutritionally improved straw in rations for finishing cattle. More recently the emphasis has been on on-farm alkali treatment of straw and the evaluation of its use in various beef systems. The work has been carried out at High

Plate 5.10 Making a cocktail treatment of straw by mixing in a feed mixer trailer.

Mowthorpe and Gleadthorpe EHFs, both basically arable, cereal growing farms with subsidiary beef enterprises utilizing limited grassland areas.

Work at High Mowthorpe

Background

At High Mowthorpe the beef enterprise is a late spring calving suckler herd of Hereford × Friesian cows with the progeny by Charolais bulls taken through to slaughter. Replacement heifers are home reared to calve at about 26 months old.

The herd size has been increased to 150 cows, sufficient to fully utilize the permanent grass on the farm with minimal conservation. What silage is produced is fed to finishing cattle and the roughage for wintering the heifer replacements and the cows comes from the arable side of the farm in the form of straw.

Traditionally straw has formed an important part of the rations for such stock in arable areas, because it is available and cheap, and because such cattle have high appetite levels relative to their feed requirements. For example a conventional winter ration for mature suckler cows at High Mowthorpe would be ad lib spring barley straw supplemented with 1.5 kg per day of barley/urea. On this basis cows can be wintered for about £77 which represents over 50% of the variable costs of suckled calf production.

Since 1977 High Mowthorpe have been looking at ways of modifying such traditional rations, first in an attempt to reduce winter feed costs and secondly because of the swing from spring barley to winter wheat production. This has gone to the extent that they no longer produce sufficient spring barley straw for our feed requirements.

It is in this context that they have looked at the value of alkali treatment of both barley and wheat straws in rations for replacement heifers, as the basis of the winter feed for mature suckler cows or, latterly, as silage replacements for finishing suckled calves.

Feeding replacement heifers on sodium hydroxide treated straw

Work on treating straw began in 1977 using sodium hydroxide which was the only contractor service for on-farm alkali treatment available at that time. Treatment was using commercially available equipment to chop straw and apply sodium hydroxide solution at a rate sufficient to provide 5% NaOH in the straw dry matter. Over four years the average improvements in digestibility were 11.5 and 9.4 units for barley and wheat straws respectively.

In each of the four years a feeding trial was carried out using seven month old Hereford × Friesian heifers starting the winter at about 185 kg. The target gain over the five month winter period was about 0.70 kg per day. The trials compared rations based on untreated barley straw and treated barley and wheat straws. The mean results are presented in Table 5.13.

Table 5.13 Performance of growing heifers on straw based diets

	Long barley straw	NaOH treated	
		barley straw	wheat straw
Daily liveweight gain (kg)			
1977 (+2kg/day barley protein)	0.43	0.52	0.52
1978-1980 (+3kg/day barley protein)	0.66	0.81	0.77
Straw dry matter intake (kg/day)			
1977	2.6	3.1	2.8
1978 – 1980	2.2	3.2	3.1

Alkali treatment led to markedly higher straw intakes and resulted in much improved liveweight gains.

The contractor system used for sodium hydroxide treatment worked well, although providing storage space for chopped, treated material posed some problems. Increases in straw digestibility were good and animal performance was excellent. The high sodium content of the feed did lead to increased water intakes and urine outputs with consequently increased bedding requirements. With small animals like this the increase was not great and was commercially acceptable. With cows, however, the increased urine production on rations based on sodium hydroxide treated straw was such that it increased bedding requirements by about 30%. It was for this reason that High Mowthorpe changed to ammonia treatment of straw for cows when this became available as a contractor service.

Feeding suckler cows on ammonia treated straw
This work began in 1981 and has generally been confined to mature cows, housed, and fed straw-based diets through months four to nine of pregnancy prior to them calving outside at grass in May/June. Treatment was by injecting aqueous ammonia at 100 litres of a 37%

solution per tonne of straw into sealed stacks of small bales, or polythene tubes filled with large round bales. Over four years the average improvements in digestibility were 9.0 and 9.4 units for wheat and barley straws respectively. These levels of increase are similar to those recorded on commercial farms by ADAS Nutrition Chemists but the range has been less. They have not experienced the very low levels of improvement measured on some commercial farms. Their recipe for treatment would be to be painstaking over ensuring a good seal, treating soon after sealing and as soon after harvest as possible (when temperatures are high) and netting to ensure that wind damage to the polythene is minimized and that the seal is maintained for at least three weeks after treatment. They have obtained better improvements in digestibility from treating damp straw, and have recent evidence suggesting that some varieties are more responsive than others.

Results of the feeding trials are summarized in Table 5.14. With the straw based diets special attention was given to mineral supplementation. The group-fed trial involved seven animals per treatment per year 1982-1984 and the individually-fed trial six animals per treatment per year 1983-1984.

Table 5.14 Performance of suckler cows on straw based diets (means of 3 years 1982-1984)

	Untreated	*Treated*	
	barley straw + 1.5 kg barley/urea	*barley straw*	*wheat straw*
1. *Group-fed trial*			
Weight loss yarding to post-calving (kg)	33.6	52.4	41.7
Calf birthweight (kg)	45.0	44.1	42.2
Straw intake (kg DM/day)	8.1	8.9	9.4
2. *Individually-fed trial*			
Weight loss yarding to post-calving (kg)	35.2	68.4	—
Calf birthweight (kg)	45.5	44.7	—
Straw intake (kg DM/day)	7.0	7.6	—

Cows in individual feeders ate less than those group-fed but in both situations alkali treatment led to enhanced intakes. As a result cows group-fed on treated barley straw or treated wheat straw alone had acceptable levels of performance. Calf birth weights were normal and weight loss over the winter was acceptable and at a level which could be recovered over the subsequent summer grazing period. Cows individually fed on treated barley straw alone had such low intakes that the winter weight loss was greater than the weight likely to be recovered at grass. However under a group feeding system mature cows have been wintered successfully for three consecutive years on rations of treated wheat straw or treated barley straw alone. This was successful despite the fact that protein supplied by the straw-only diets was below theoretical requirements for such animals. It is important to emphasize that the feeding of treated straw alone cannot be advocated for young cows which are still growing, cows which start the winter in poor conditions or cows in lactation.

Compared with the conventional ration of untreated straw and 1.5 kg/day of barley/urea (95% barley, 5% urea giving 23.3% CP) we would not consider feeding treated barley straw as an economic proposition in our circumstances – the cows will eat large quantities of untreated barley straw anyway and money is generally better spent on supplementation than treatment. However, on a farm like High Mowthorpe, where wheat straw can be costed into the system at no

Plate 5.11 Adding molasses and Granstock to make a cocktail of treated straw (also using solid caustic soda) in a feed mixer trailer.

more than cost of baling and handling (it would otherwise be burnt), treatment of wheat straw with ammonia can be justified. With wheat straw at £8 per tonne and treatment at £21 per tonne feeding treated wheat straw alone can save up to £6 per cow over the winter period.

Feeding finishing cattle on ammonia treated straw
High Mowthorpe have done one trial on this using weaned suckled calves intended for finishing in yards at about a year old. They fed ad lib untreated spring barley straw supplemented with 4.5 kg per day of barley/protein (11% soya) as the control. Compared with this they fed ad lib ammonia treated barley straw supplemented with 3.0, 3.5 or 4.0 kg per day of barley/protein (15, 11 and 7% soya respectively). Results for a 113 day feeding period are given in Table 5.15.

Table 5.15 Performance of suckled calves on straw based diets

	Untreated straw 4.5 kg supp.	Treated straw 3.0 kg supp.	3.5 kg supp	4.0 kg supp.
Liveweight gain (kg/day)	0.81	0.71	0.77	0.82
Straw dry matter intake (kg/day)	1.9	2.8	2.4	2.4
Total dry matter intake (kg/day)	5.7	5.4	5.4	5.8

More straw was eaten by animals fed treated straw but levels of supplementation were lower so that total intakes were similar on all treatments. Animals fed treated straw and 4.0 kg supplement performed similarly to those fed untreated straw and 4.5 kg supplement, treatment allowing an apparent saving of 0.5 kg supplement per day. However, when the cost of treatment and the increased intake of treated straw are taken into account the daily feed costs of both treatments were the same (67p). In addition it should be pointed out that this trial terminated after 113 days because none of the treatments was producing sufficiently high growth rates for the cattle to fatten.

Work at Gleadthorpe

Background
At Gleadthorpe the area of grassland has been progressively reduced over recent years and there has been a deliberate decision to stop

making silage because of costs and clashes with arable workloads. Winter feeding is now based on purpose grown forage crops, mainly fodder beet, or arable by-products like beet tops, potatoes and straw, which at Gleadthorpe generally means spring barley straw.

Dairy-bred calves or store cattle are purchased and either wintered on forage crops and finished on grass, or are finished indoors on cereal based rations. Is is the latter enterprise for which ammonia treated straw has been investigated, either to replace the limited amount of untreated straw fed and give better performance, or to cheapen the overall ration by replacing part of the cereal component as well.

Mr Bastiman's general conclusions were:

'Under EHF conditions alkali treatment of straw has proved relatively reliable and over large batches of straw has led to an average increase in digestibility of about 9 units. Treated straw has proved palatable, and one of the major benefits of treatment has been increased straw intakes. However it should not be forgotten that even after a 9 unit increase in digestibility treated straw is still no better than moderate hay.

'Feeding beef cattle on treated straw has produced no surprises and levels of performance obtained have been similar to what would have been predicted from calculations based on the measured ME intakes.

'However treatment of barley straw has not proved economic. Untreated barley straw is eaten quite readily and in all the circumstances tested the benefits from alkali treatment could have been obtained more cheaply by increased supplementation of untreated straw. The situation with wheat straw is different. It is a cheaper commodity to start with (in many cases regarded as a liability rather than an asset) and is generally unpalatable. Treatment of wheat straw can turn an otherwise waste product into an acceptable feed. Even so the situations where it could be economically justified are likely to be limited to where it is grown on the farm on which it is fed, to where it represents a high proportion of the diet (i.e. where low levels of performance are required) and to where it is being fed to animals with high intake potential. In our view this limits its use to overwintering dry suckler cows or as part of a store ration. Under circumstances where higher levels of performance are required the level of inclusion of straw and therefore the potential benefit to treatment are severely limited.'

Plate 5.12 On the left is a molasses tank while on the right is a safe dispenser for solid caustic soda for the cocktail treatment of straw.

COCKTAIL TREATED STRAW – PRACTICAL DIETS

In 1979/1980 the author began to work with Mr Ian Leonard of Leonard Farms in Norfolk on the 'cocktail' treatment of straw for beef diets. The objective was to feed 40-50% of straw, without grain, and achieve over 1 kg of liveweight gain a day. The process used is still in use and is outlined below.

Procedure

The straw is chopped and put in a complete diet feeder. The 'Pearl' caustic is added and mixed for about 2 min and then the urea is added. The molasses are added next and then the rest of the ration.

After 3-5 min mixing (in the type of mixer used by Mr Leonard, a Farmhand 280), the diet is ready for feeding.

Straw

Chopped and treated with:
 — solid 'Pearl' caustic soda, at 5% rate
 — Granstock (urea) at 30 l/tonne straw
 — molasses at 55 kg/100 kg straw.

Safety precautions

The procedure is relatively very safe. Gloves and goggles must be worn when handling the caustic soda which should pass a 60 mm (or 0.25 in) sieve to remove lumps. Urea, molasses and other ingredients are best kept away from the skin and eyes.

Diets at Leonard Farms

Three diets were fed to beef cattle in the winter of 1980/1981. The details of these diets and their calculated results are outlined below (Table 5.16).

On a theoretical basis, the three rations fed would have given the performance figures outlined below in Table 5.17.

Table 5.16 Diets at Leonard Farms

1. Fresh kale	1300 kg	
Seeds chaff	350	DM = 42.8%
Grass silage	900	ME = 11.1 MJ/kg DM
Dried beet pulp	750	DCP = 75 g/kg DM
2. Kale silage	750 kg	
Seeds chaff	350	DM = 44.2%
Grass silage	1500	ME = 10.4 MJ/kg DM
Dried beet pulp	500	DCP = 80 g/kg DM
3. Barley straw	500 kg	
Dried beet pulp	400	DM = 61.5%
Grass silage	390	ME = 10.5 MJ/kg DM
Sodium hydroxide prills	25	DCP = 85 g/kg DM
Urea	30 litre/t straw	
Molasses	55 kg/100 kg straw	

Taking theoretical figures for dry matter intake (DMI):

at 350 kg liveweight = 8.5 kg DMI

400 kg liveweight = 9.4 kg DMI

and   500 kg liveweight = 10.7 kg DMI

the complete diets (1), (2) and (3) would support the following gains (Table 5.17).

Table 5.17 Expected results at Leonard Farms

Liveweight (kg)	Diet	DMI (kg) estimated	Nutrients in diet ME (MJ)	DCP (g)	Expected daily gain (kg/day)
350	1	8.5	95	638	1.2 (actual)
400	1	9.4	104	705	1.2 (actual)
	2	9.4	98	742	1.1
	3	9.4	99	800	1.1
500	2	10.7	111	845	1.1 (target)
	3	10.7	112	910	1.1 (1.7 actual)

In fact, diet No. 3 produced a liveweight gain on the first trial of 1.7 kg/day. This may have been due in part to higher dry matter intake and there is also some evidence that the *in vivo* performance of stock on caustic treated straw is better than the *in vitro*, or test tube, results would suggest, i.e. the computed ME figure under-estimates the real feeding value of the treated material. In practice, this farm has been able to repeat liveweight gains consistently in the region of 1.3-1.5 kg per animal per day. This level of gain has been achieved or exceeded on a large number of stock, over several years on a number of farms.

The analysis of the diet No. 3 was:

Dry matter	44%	
Crude protein	12.30	
Ether extract	1.85	
Crude fibre	20.04	several diets were actually used
Total ash	12.35	
In vitro DOMD	62.80	
ME (estimated)	10.05	

On the initial trial the diet was fed for eight weeks. The only problem with the stock was increased urine production. This was kept within reasonable limits by feeding salt-free minerals. This minimizes the excess sodium that the animals have to get rid of. The diet is mixed in a feed mixer wagon and there is no delay after treatment before feeding. The procedure is, therefore, simple and uncomplicated.

Plate 5.13 Solid caustic soda produced as 'Pearl' by ICI. There is very little dust in this sample of material which makes it much safer to use.

Table 5.18 Costings of cocktail treatment at Leonard Farms (1981 prices) (cost per load)

	Amount per load (kg)	Cost per tonne (£)	Cost per load (£)
Diet No. 2 (untreated straw)			
Kale silage	750	4	3.00
Seeds chaff	350	55	19.25
Grass silage	1500	15	22.50
Dried beet pulp	500	80	40.00
	3100		84.75
Diet No. 3 (straw treated with a 'cocktail')			
Ground barley straw	500	15	7.50
Dried beet pulp	400	80	32.00
Grass silage	390	15	5.85
Sodium hydroxide prills	25	235	5.87
Urea (Granstock)	30 litres/t	65/drum	4.75
Molasses	55 kg/100 st.	85	4.67
	1620		60.64

At dry matter 61.5% cost per kg dry weight is 6.08 pence. On the actual figures for dry matter intake at 10.7 kg per day on diet No. 2, and 9.92 kg per day for diet No. 3, the costs would be as in Table 5.19. It is thought that the treated diet could be produced with lower costs.

Table 5.19 Costs at Leonard Farms (1981)

	Diet No. 2	Diet No. 3 Initial trial at 1.7 kg/day l.w.g.	Diet No. 3 Commercial production @ 1.5 kg/day l.w.g.	Diet No.3 at 1.3 kg/day l.w.g.
Cost per day	22.126	60.31	60.31	60.31
Cost per kg liveweight gain (l.w.g.)	60.11	35.38	40.21	46.39

Notes
1. In commercial practice after this trial, consistent liveweight gains of 1.3-1.5 kg/day have been achieved.
2. The procedure with some makes of feed mixer trailer has, as a result of experience, been modified. Some machines need longer mixing times. Generally 5 min should be the minimum and, in some cases, 10 min. The test is to observe a dramatic change to a golden yellow colour in the straw and a significant reduction (up to 50%) in volume.

SUPPLEMENTATION

It is known that straw feeding involves the digestion system in different processes and a different mineral/vitamin supplement should be used.

With sodium hydroxide treatment it is possible, in theory, to overload the system. If sodium hydroxide were applied at the 5% of dry matter level to the straw and if this were fed at 100% of the diet, it is possible to approach a level of 2% of the diet as sodium. At such a level it is possible to affect nerve function. Metabolic profiles carried out by the author did not actually observe these results but it is, in theory, possible. In practice, it is unlikely that such a figure of 100% would be fed. Clearly, however, efforts should be made to limit the effects of sodium to the diet.

Firstly, the mineral/vitamin supplement should be free of common salt (sodium chloride). This will make it much more expensive per lb or kg but, of course, much less will be fed. Overall it will not be more expensive. This supplement should be designed for the job and will contain more phosphorus, sulphur, magnesium, copper, selenium and Vitamin E.

Plate 5.14 The author carried out trials in Essex in the mid 1970s to treat straw with anhydrous ammonia.

Secondly, stock must have free access to clean water at drinkable temperature. If there is an excess sodium problem, the stock will drink more and self-correct the excess. Restriction of water intake or failure to eliminate the common salt from the mineral supplement in processed diets containing high proportions of caustic treated straw may result in problems, particularly to the kidneys.

ACTION PLAN — METHOD CHOICE

Ammonia

 (a) For trials – possibly use stack method ⎫
 (b) If capital available – ovens   ⎬ with anhydrous ammonia
 (c) Otherwise – aqueous treatment.   ⎭

Caustic soda

 (a) With forage harvester possible but safety problems.
 (b) With purpose-built treatment machine using liquid caustic.
 (c) For cocktails, use solid caustic soda and feed mixer trailer.

ACTION PLAN — FEEDING

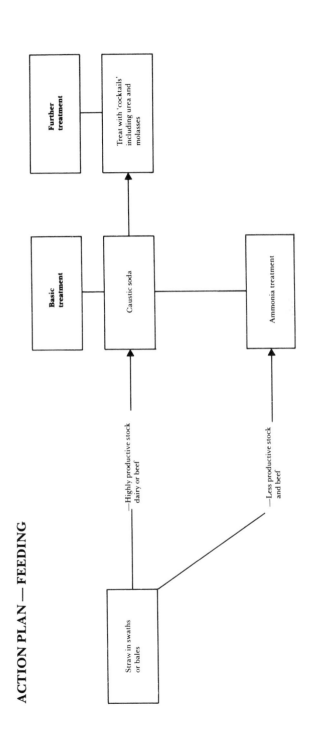

Basic treatment

Further treatment

Straw in swaths or bales

—Highly productive stock dairy or beef

—Less productive stock and beef

Caustic soda

Ammonia treatment

Treat with 'cocktails' including urea and molasses

6

Straw as a fuel

STRAW AS A FUEL — ITS BASIC VALUE AND CHARACTERISTICS

Nobody would argue that straw is not a potentially useful source of heat. The questions are really:

- How much heat is there in straw?
- What mechanics are involved in its use as a fuel?
- How, when and where can the fuel be cashed from an economic point of view?

Straw should be put in context with other fuels as a source of heat (Table 6.1).

Table 6.1 Straw as a fuel compared to other fuels

	MJ/kg	BTU/kg	kWh/kg
Straw	16	15 165	4.45
Elm wood	16	15 165	4.45
Peat	17	16 113	5.72
Good coal	30	28 434	8.35
Oil	40	37 912	11.13

Clearly straw has some value especially when put alongside costings. Table 6.2 shows basic costs and Table 6.3 costings related to likely efficiencies of use.

It will be noted in Table 6.3 that the figures for straw are based on a cost of £20 per tonne and given at two efficiencies of 50 and 80%. In practice, different furnaces vary widely in their efficiencies of burning. Most furnaces also depend on the moisture content of fuel.

Table 6.2 Comparison of energy sources at 100% efficiency of use

Fuel	Bulk wood and straw	Natural gas	High grade coal	Oil	Electricity	
					Offpeak storage 2.5 p/unit	Normal for heating and domestic hot water 4.9 p/unit
Comparison cost of 'raw material'	£20/tonne own-grown fuel can be costed lower	35.2 p/therm	£120/tonne	20 p/litre		
Cost per 100 000 BTUs	25 p	35 p	42 p	70 p	73 p	144 p
Cost per kWh	0.45 p	1.21 p	1.44 p	2.39 p	2.02 p	4.9 p

Table 6.3 (from Table 6.2)

Fuel	Cost/kWh (p)	Efficiency of use (%)	Cost per effective kWh (p)
Straw	0.45	50	0.90
Wood	0.45	80	0.56
Gas	1.21	70	0.64
Coal	1.44	80	1.51
Oil	2.39	70	2.06
Electricity	4.9	100	4.90
Electricity (off peak)	2.02	100	2.02

Plate 6.1 A 'turbo' burner, burning a controlled supply of chopped straw in excess of air. The combustion chamber is shown on the right hand side of the main boiler unit.

Moisture content of straw

Figure 6.1 shows that the usefulness of the heat in the straw, in most furnaces, is inversely related to moisture content.

All dry straw has about the same net calorific value when perfectly dry, viz. approximately 16 MJ/kg (capable of providing about 4.5 kWh of heat). However, when moisture is present, as it inevitably is, a proportion of the heat is absorbed in vaporizing this moisture, and thus less is available for useful purposes. 'Dry' straw may contain up to 15% moisture and this reduces the available heat to about 14 MJ/kg (4 kWh). The useful heat developed is inversely proportional to the moisture content (as shown in Fig. 6.1) and hence in terms of effective utilization of the straw (not the least, the task of manhandling it) it pays to burn the driest straw that is available.

In fact, while such a statement is generally true, we have now got heat exchangers for furnaces which make the overall efficiency independent of moisture content. This sounds like an engineer's dream but it does, in practice, work (see section on Equipment, p. 200).

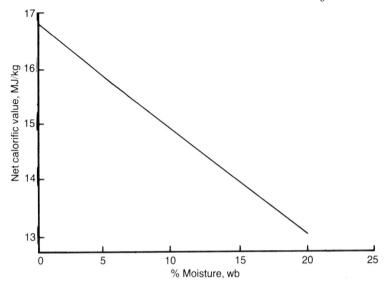

Fig. 6.1 Heating value of straw vs. moisture content.

On-farm costs of straw

The figure of £20 per tonne used in Table 6.2 above is, in fact, a commonly used guide to straw cost. There is however a wide range of costs associated with moving straw from the field to it being used as a fuel. For example, straw has an agronomic value that must be met if it is removed and baled; the baling technique and area baled will substantially affect costs. There are also costs associated with handling and on-farm transport, storage, transport off-farm, and any further processing required (e.g. briquetting, or chopping) prior to combustion. Silsoe College have collected information on the acquisition cost of straw through a farmers' survey, and have developed extensive system models to enable the other operations to be costed. Some of these results are presented in Table 6.4 which shows costs that have been extracted for various fuel uses. Straw for on-farm use as fuel is shown as £18/tonne but some farmers already baling straw may assume a zero cost for straw in the swath if they themselves are to use it. Also, storage in disused buildings could reduce storage costs significantly. Thus a minimum figure of £10/tonne for straw used on the farm is appropriate.

Table 6.4 The cost of straw as a fuel

	Cost range (£ tonne)	Typical cost (£ tonne)
1. Price of straw, ex-swath	0–12	5
2. Baling, handling and on-farm transport	7–14	9
3. Storage on-farm (including losses)	3–6	4
4. Transport		
20 km round trip	3-10	4
50 km round trip	4-13	6
320 km round trip	6-16	10
Totals (a) On-farm use (1+2+3)		18
(b) Rural industrial use		
(1+2+3+4(20 km))		22
(c) Industrial use		
(1+2+3+4(50-320 km))		24-28

Larry Martindale of the Energy Technology Support Unit at Harwell gave an analysis of the situation in *Straw Disposal and Utilisation* published by MAFF in July 1984.

'It is evident that there is technology available that is capable of processing and combusting straw efficiently and over a wide range of outputs. The smallest boilers are already being used on-farm, while larger installations have been demonstrated in Denmark and are to be demonstrated in the United Kingdom over the next few years. The best opportunities will be where high rates of utilization can be achieved. Straw fired boilers, furnaces and preparation plant are expensive to install and operate, and so high utilization will enable straw to compete more effectively with higher priced fuels such as coal and oil.

'Use of straw as fuel on-farm is already cost-effective, and there are about 7700 boilers in the United Kingdom that are at least in part fired by straw. These consume around 166 000 tonne of straw (0.09 Mtce – million tonnes coal equivalent) per annum. The majority are simple whole-bale burning boilers used for space heating (farmhouses and animal houses) or crop drying, typically with outputs up to 0.17 MW. Increasingly, for applications with a more critical heat cycle, for example, glasshouse heating, automatically stoked boilers are being used; the largest so far installed in the United Kingdom has a rated output of 1.1 MW.

'In the short term, the most significant outlet for straw as fuel is for on-farm applications, such as have been outlined above. Silsoe College have analysed the straw distribution with respect to potential users, and report that the maximum utilisation of straw on-farms could be 2.1 million tonne (1.1 Mtce) per annum. However, it is unlikely that the full potential will be exploited, and the study suggests that use is likely to reach only 1 million tonne (0.5 Mtce) per annum by the year 2000.

'For use in rural industry, straw is currently only marginally more attractive than coal, although the economics versus oil are generally much better. However, haulage distances must be short since even modest increases in straw acquisition cost can negate its economic advantage. Coal prices have risen during the last few years relative to straw prices and the trend seems destined to continue; the use of straw will thus increasingly be attractive for rural industrial applications.'

Alternative wastes

Straw, then, appears to be a potentially useful source of heat. It is relatively cheap per unit of heat supplied, but it has a handling and storage problem which is of practical significance.

Plate 6.2 An electrically driven tub grinder can be used as an automatic feed for supplying chopped straw to a boiler.

Table 6.5 Types of waste fuels and possibilities for use in cement-making

Type	Example	Approximate fuel value (kcal/kg)	Possible methods of use	
			Finely ground	*As small lumps or grit*
Brittle, low ash content	Charcoal fines Rubber char Petroleum cokes Some brown or lignite coals (30% moisture)	5000 – 7000 5000 – 7000 7300 – 7500 4000 – 4500	Fired in burning zone of rotary kilns	– to riser pipe of s.p. kilns, up to about 100 kcal/kg clinker. – to precalciners, up to about 400 kcal/kg clinker.
Brittle, high ash content	Colliery minestone, shale or washery tailings High carbon pulverized fuel ash (PFA) from inefficient coal-fired electric power stations Oil shales	500 – 4000 1000 500 – 2500	Fired in burning zone of rotary kilns, up to about 10% total ash on clinker To riser pipe of s.p. kilns, up to about 100 kcal/kg clinker or up to about 20% ash on clinker To precalciners, up to about 500 kcal/kg clinker or up to about 30% ash on clinker	– ground in with dry raw meal for s.p. kilns, up to about 70 kcal/kg clinker. – ground in with kiln feed slurry wet process kilns, up to about 300 kcal/kg clinker. – burned in separate fluid bed combustor/hot gas generator for raw materials drying.
Tough, non-brittle or large size	Domestic refuse Municipal garbage Oil palm shells (8% moisture) Peat (10% moisture) Vehicle rubber tyres Acid battery cases Wood chips	1500 – 2500 4800 4000 – 4500 5000 – 7000 3000	Disintegrate or shred and where required partially dry: fire into front end of rotary kilns up to about 10% ash on clinker to riser pipe of s.p. kilns, up to 100 kcal/kg clinker or up to about 20% ash on clinker to precalciners, up to about 500 kcal/kg clinker or up to about 10% ash on clinker burned in separate fluid bed combustor/hot gas generator for raw materials drying	

		As received:	
Tough, non-brittle but small size and/ or low density	Rice husks	3500	fire into front end of rotary kilns, up to about 10% on clinker to riser pipe of s.p. kilns, up to about 100 kcal/kg or up to about 20% ash on clinker
	Saw dust	3500	to precalciners, up to about 500 kcal/kg clinker or up to about 30% total ash clinker
	Chopped straw	3500	
Fluid	Waste oils, oil re-refining residues. Waste organic chemicals, acid tar	5000 – 10000	Fire as liquid fuel to front end of rotary kilns, to back end of s.p. or precalciner kilns, or for raw materials drying

General note. In general, high volatile matter content materials are best fired to the front-end of rotary kilns, whilst low volatile matter materials are best fired to the back end of s.p. precalciner kilns.

Mr Coomaraswamy of the Blue Circle Group gave a paper at the 1982 Oxford Straw Conference and gave the alternative data in Table 6.5.

Conclusion

Straw is a potentially useful and economic fuel. It is present on farms and often needs to be moved or disposed of. Farm staff understand it as a material in terms of its handling and they have the equipment to do it. The economics look attractive if capital costs of furnaces can be reasonably paid for out of savings in fuel costs.

Straw has one potentially over-riding advantage. It is an annually renewable resource available on the farm, under farm control. The farm is not dependent on an external source of supply.

The handling problem

The real problem with straw furnaces is the handling of straw both into the furnace feeding system and within the furnace itself. Straw is low value, high bulk material.

Table 6.6 Density of straw compared with other fuels

	kg/m^3
Coal briquettes	1200
Straw pellets	600
High density bale	240
Fine ground straw	200
Standard bales	100
Big bales	50 – 100
Forage harvested straw	50

As Table 6.6 shows, straw is not a dense material. It is not dense enough, in the bale, to form economic loads on lorries and therefore, it needs to be compressed to be transported efficiently. For an economic load on a lorry a density of at least 260 kg/m^3 is looked for. Straw pellets would provide that. There is a problem. Straw baling uses up energy and will use (in diesel to drive the baler tractor) about 2% of the energy available in the straw itself just to bale it up.

However, pelleting takes a lot more power and about 40% of the available energy in the straw would have to be spent driving the pelleter. By the time the straw pellets had been handled, transported

Plate 6.3 An alternative automatic feed is to place conventional bales on a conveyor (left of the photograph) for feeding a thermostatically controlled feeding device at the back of the boiler unit which burns a controlled supply of straw in excess of air.

and got to the point of use a lot more energy would be used. The plain fact is that pelleting is just not on at present – it may never be. The conclusion must be that the most economic use of straw is going to be on or very close to the farm where it is produced.

Chapter 4 goes into farm handling in some detail. It is useful here to look at the implications of handling for feeding straw to furnaces.

Doug Smith of Vicon Limited, at the 1982 Oxford Straw Conference gave the figures in Table 6.7 and 6.8. As these tables and Chapter 4 show, handling can be streamlined and there is strong reason to go to larger than conventional sized bales. This gives the furnace engineer a problem in feeding the furnace. There are solutions (see below on equipment) but these tend to have a high capital cost.

The Combustion Process

I.E. Smith of the Cranfield Institute of Technology described the combustion process at the Oxford Conference in 1980.

Table 6.7 Relative bale densities

System	Dimensions (m)	Volume (m³)	Weight (kg)	Density (kg/m³)
Conventional	0.36 × 0.46 × 0.95	0.157	16	100
Conventional	0.36 × 0.46 × 1.15	0.19	25	130
Round	1.20 × 1.80 dia.	3.05	300	100
Large square	1.20 × 1.27 × 2.5	3.80	550	144
VICON HP 1600	1.60 × 0.70 × 1.20	1.35	200 – 220	150 – 160

Typical bale weights on silage would be 600 kg, and on hay 350 kg.

Table 6.8 Typical payloads using 12 m flat bed trailers

System	Bale dimensions (m)	Bale weight (kg)	No. per load	Weight (t)
Conventional	0.36 × 0.46 × 0.95	16	512	7.0
Conventional	0.36 × 0.46 × 1.15	25	400	10.0
Round	1.20 × 1.50 dia.	210	23	4.8
Large square	1.20 × 1.27 × 2.50	550	20	11.0
Vicon HP 1600	1.60 × 0.70 × 1.20	200	64	12.8

Burning a bale

All solid fuels, coal, wood and straw contain carbon, hydrogen and oxygen in varying proportions. If they are fed into the grate of a boiler or furnace the combustion takes place in two stages. The first is a volatilization process when hydrocarbons literally evaporate, mix with atmospheric oxygen and then burn in the gaseous phase. This stage having been completed, the residue is simply a 'char' consisting principally of carbon which cannot evaporate, but has to rely on the diffusion of oxygen to its surface when it is oxidized to carbon dioxide (CO_2) or carbon monoxide (CO).

The volatilization stage is governed by the diffusion of heat that is transferred from the combustion gases and hot burning bodies in the vicinity and more importantly to the resistance of the particle or body that is actually burning to the transfer of heat into its interior. This provides a clue as to why the combustion of relatively loosely packed straw is quite different from that of lumps of coal or logs. Large particles require a long time for heat to diffuse into their interior and hence a thermal gradient is established within them as they burn in a furnace, which persists until they are almost completely combusted. Thus, volatilization and char combustion take place simultaneously, as is shown in Fig. 6.2. With straw, even in bale form, the diffusion of heat, and hence volatilization, is extremely rapid on account of the

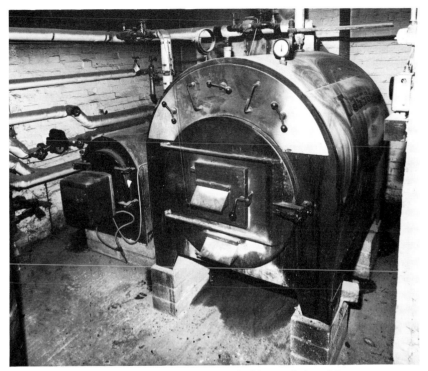

Plate 6.4 A Passat straw burner used in parallel with an existing oil burner. The thermostat will cut in the oil if necessary.

small stem size, and so the first process occurs very quickly indeed. Hence, after feeding a straw bale into a furnace, the combustion of the volatile materials may be completed in a comparatively short period of time, after which the much slower combustion of the char has to follow.

In most older furnaces, that is what happens. A bale is burned as a whole bale and the rate of burning is controlled by controlling the air supply, i.e. excess of straw is burned in a limited air supply. Efficiencies of burners of this type rarely achieve more than 50% even with very dry straw.

There is an alternative and that is to control heat production by controlling the supply of straw, i.e. a limited supply of straw is burned in excess air. In this way, efficiencies will rise to 80 or even 85%.

To put the efficiency levels into economic perspective leaves straw

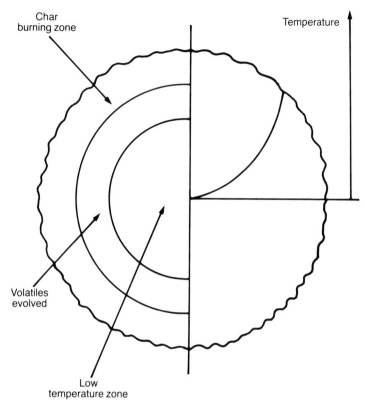

Fig. 6.2 The combustion of solid fuel.

with a value of a sufficient level to leave a good margin on running costs to pay off capital investment in a furnace.

From Table 6.1

Basic value of straw	4.45 kWh/kg
Burned at 80% efficiency	3.56 kWh/kg
but 1 kWh of oil costs	2.99 p
therefore:	

1 kg straw replacing oil would be worth 3.56 × 2.99 = 10.64 p therefore 1 tonne straw is worth £106

To put it another way, with high efficiency burning, a small bale at about 30 lb replaces a gallon of heating oil or 3 kg straw replaces 1 litre of oil.

Ash from furnaces

Compared with wood which has 0.5–5% ash, cereal straw is relatively high in ash. Wheat straw gives about 6% ash and has low sulphur content. With efficient burning, flue gas emissions are comparable to oil. With the excess air burners using chopped straw, they can even comply with strict German clear air legislation.

The ash is mainly P and K with useful Mg content. The level of silicates, especially in barley straw, is quite high. Rice ash is even higher. These high silicate ashes make quite useful cements for mortar or concrete. Otherwise the ash has useful manurial value.

THE EQUIPMENT AVAILABLE

Single pass boilers (air to water heat transfer)
Early designs of furnace were really rather crude incinerators with a steel chamber in which a straw bale was burned. There was a door at one end, a flue at the other and a water jacket around the cylinder to produce a supply of hot water. The rate of burn was controlled by an air flap in the door. Further development produced two air supply vents in the door; the lower one produced the initial 'flair' and the upper one supplied air to mix in a turbulent flow with the volatiles giving the main burn and heat supply in the upper part of the combustion chamber. The design is still the simplest and lowest

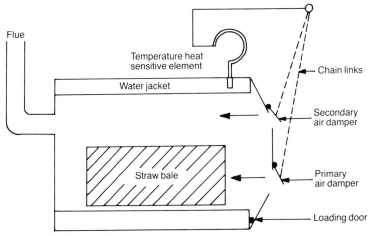

Fig. 6.3 Single pass burner.

available capital cost (Fig. 6.3). These boilers are manually stoked and have a burning efficiency of about 50% at best. Loss of heat from the outside of the boiler is also high. Some of the big bale burners (e.g. Farm 2000 or Scanfield) can reach 60+ % efficiency because they use a forced draught.

Because of the low efficiency, ash production is relatively high and frequent removal (every 1-3 days) is likely to be necessary. It is also likely that flue gases will be visible and may contain smuts particularly at low flue gas temperatures which follow the use of damp fuel.

Triple pass furnaces (air to water heat transfer)

A current development on this theme with minimal moving parts is the triple pass boiler with a more complicated flue. The extra surface area for transferring the heat from the flue gases to the water raises potential efficiency to possibly 60%, maybe a little higher.

The flue is relatively constrained and will need regular cleaning. With damp fuels giving low temperature flue gases, this may be quite

Plate 6.5 Simple incinerator type boilers have low efficiency but are low cost and have few problems provided dry straw is used.

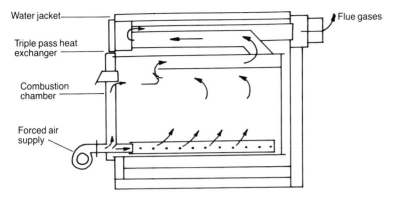

Water jacket

Triple pass heat exchanger

Combustion chamber

Forced air supply

Flue gases

Fig. 6.4 Triple pass furnace.

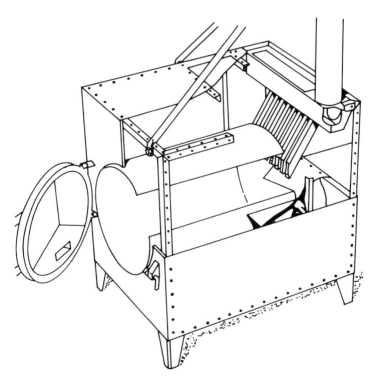

Fig. 6.5 An air-to-air type incinerator.

a problem. At low flue temperatures, it is likely that deposits of heavy hydrocarbon gums and lacquers will be deposited in the flue. The easy way out is to ensure the use of dry fuels.

As with the single pass boiler, the triple pass design relies on burning excess of straw (a whole bale) in a controlled (limited) supply of air. The results on ash and flue gases will be the same (Fig. 6.4).

Incinerators (air-to-air type)

The basic idea of a furnace without a water jacket is that it is simpler to build, potentially less expensive and independent of a water supply, so it is portable. As with single pass water boilers, efficiency would be low but a fan air supply can lift efficiency a little. The heat produced is in air which is drawn over the combustion chamber by an external fan as, for example, built into an on-floor grain drying system. The approach, then, has its advantages as a simple system in grain drying (Fig. 6.5).

Excess air furnaces

A high level of burning efficiency, up to 80 or even 85% is possible if a limited and controlled supply of straw is burned in excess air. This allows a thorough 'char' and almost complete burning of the volatiles; hence the high level of efficiency.

There are a number of alternative ways of feeding the straw to the furnaces. The rate of feed is controlled by a thermostat.

Straw can be shredded or chopped and fed by auger to the base of the burning grate. A small fire would then burn in the grate which would be fed with a forced supply of excess air. An alternative is suspension burning where a relatively finely chopped sample of straw is fed above the grate and is ignited by hot air. The straw burns partially or completely while still suspended in the air supply. A high level of efficiency of burning is possible using these methods of combustion (Fig. 6.6).

Rate of straw use

Generally single pass boilers will burn about 5 kg of dry straw to replace 1 litre of oil. Triple pass furnaces will burn about 4 kg of dry straw and 'turbo' burners (using excess air) about 3 kg to do the same thing.

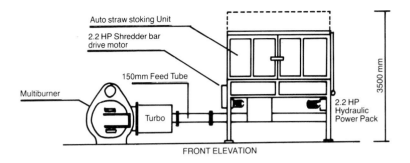

Auto straw stoking Unit

2.2 HP Shredder bar drive motor

150mm Feed Tube

Multiburner

Turbo

3500 mm

2.2 HP Hydraulic Power Pack

FRONT ELEVATION

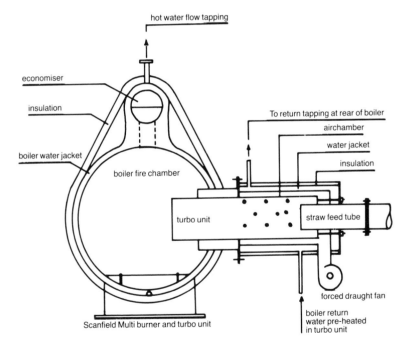

hot water flow tapping

economiser

insulation

boiler water jacket

boiler fire chamber

To return tapping at rear of boiler

airchamber

water jacket

insulation

turbo unit

straw feed tube

forced draught fan

boiler return water pre-heated in turbo unit

Scanfield Multi burner and turbo unit

Fig. 6.6 Automatic feed straw burner supplying chopped straw, controlled by a thermostat. Burning of straw is in excess of air in the 'turbo unit'.

Independence of fuel moisture content

All the above designs and performance figures have been based on the assumption of 'dry' fuel with a limit of preferably 15% moisture. In conventional furnaces a moisture content of 23% would, in theory, prevent the emission of useful heat and, in practice, the furnace would probably go out anyway.

Clearly, 'water does not burn very well' is an expression that is very useful in practical management of conventional furnace systems. However, there is an ingenious way round the problem developed by Henrik Have of the Royal Veterinary and Agricultural University of Denmark. The idea depended on the observation that the drop in temperature of flue gases was very marked as moisture content of fuel rose. The mechanism involved was that to evaporate water, latent heat was involved in the change in state of the water from liquid to gas. The latent heat dramatically reduced flue gas temperatures. If the latent heat could be trapped, it could be used. The key was to 'scrub' the flue gases in a water scrubber. The moisture present in the gases as a vapour is then changed back from a gas to a liquid and the latent heat released, so raising the temperature of the scrubber water. In theory, at least, this attachment to the flue of a furnace makes the efficiency independent of fuel moisture content.

The development work in Denmark concentrated on a furnace for burning animal slurries as fuel but the idea could, in practice, be used

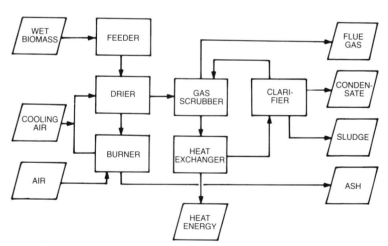

Fig. 6.7 Flow chart showing system for heat extraction from biomass by combustion (Patent pending).

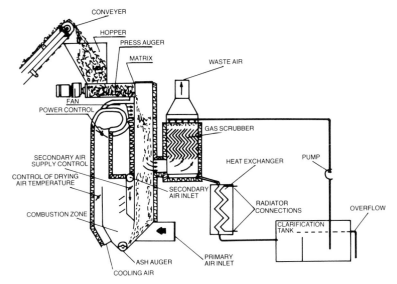

Fig. 6.8 Suggested design of plant for combustion of animal manure.

with a conventional furnace. Figures 6.7 and 6.8 show the original Danish furnace for burning wet slurry and the built-in drying mechanism which allows the fire to stay alight.

Figures 6.9–6.13 show furnace types and feeding systems (courtesy of Henrik Have, University of Copenhagen).

Financial considerations

Single and triple pass boilers will have a capital cost in the £1000–£5000 bracket. Bearing in mind that the average farmhouse will spend about £2000 on its oil heating bill, the capital cost could easily be discounted. However such boilers will have to be stoked at least once and maybe twice a day.

Automatic feed boilers will cost £10 000 upwards and short season use may well not be enough to justify the cost. However, hot water can be piped long distances in installed, underground pipes and a large installation could heat several houses and dry grain or heat animal housing.

Larry Martindale (in *Straw Disposal and Utilisation* published by MAFF in July 1984) drew a graph to show pay back periods (Fig. 6.14).

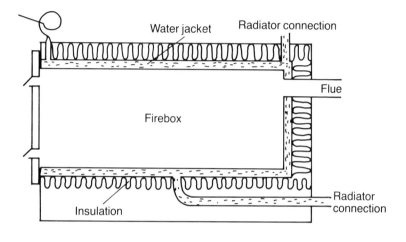

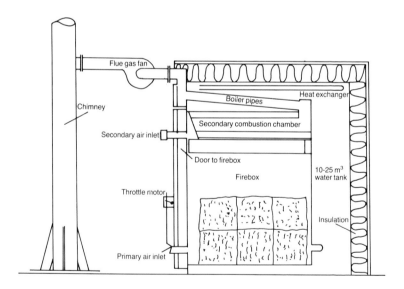

Fig. 6.9 Updraft, fixed bed straw furnaces.
Top: **for small bales (Passat).**
Bottom: **for big bales with heat accumulating water tank incorporated in the furnace (BB-energi).**

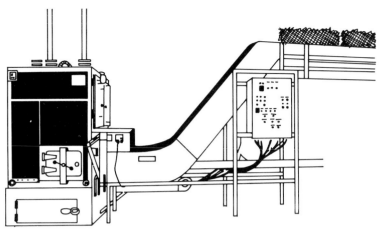

Fig. 6.10 Updraft, fixed bed straw furnace (left) with automatic stoking system (right) which feed one bale at a time (LIN-KA).

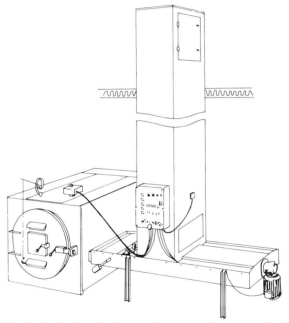

Fig. 6.11 Updraft, fixed bed straw furnace with automatic stoking system which feeds part of a small bale at at time. The bales are fed through the door in the top of the straw hopper. In the bottom they are cut and pushed into the firebox by a horizontally-moving piston with knife (Holstelro).

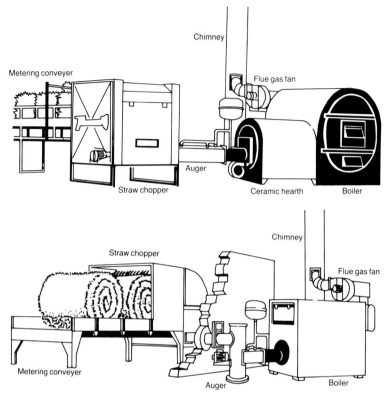

Fig. 6.12 Furnaces fed by automatic stoker system and straw chopper (Passat).
Top: **small bales;** *Bottom:* **small or big bales.**

'The economics of installing and operating oil, coal and straw fired boilers for rural-industrial and industrial applications are shown in Fig. 6.14. Simple payback periods have been calculated for a wide range of boiler outputs from 0.3 to 14.6 MW. Industrial experience suggests that utilisation rates increase with boiler capacity, so rates are on an increasing scale from 18 to 70%. Similarly coal and oil prices vary according to the level of consumption, and larger boilers will generally use heavy fuel oil rather than gas-oil. Since the delivery distance for straw will increase with increasing boiler capacity two delivered straw prices are used; for boilers of up to 4.4 MW capacity a price of £22/tonne is used, while for larger boilers a price of £26/tonne is used.

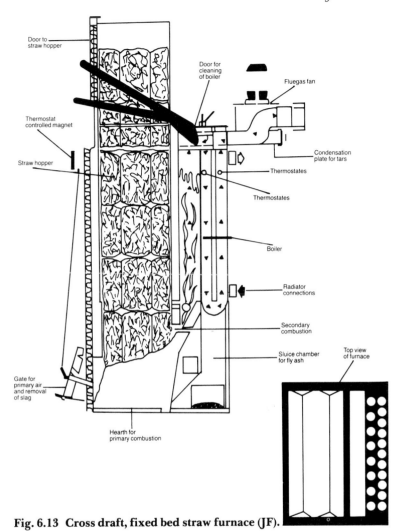

Fig. 6.13 Cross draft, fixed bed straw furnace (JF).

'Clearly, for the smaller boilers (0.3 MW) straw is cost effective with respect to coal, but not with respect to oil. For larger scale boilers (around 1.5 MW) straw becomes cost effective with respect to oil, offering payback periods of 5 to 10 years; the payback period for straw versus coal is 3 to 12 years. For boilers of 4.4 MW, straw is no longer cost effective with respect to coal (with minimum payback periods of 14 years) although it becomes increasingly attractive compared to oil, with payback periods of 3 to 4 years with respect to gas-oil or 6 to 8 years with respect to heavy fuel oil. Although straw can still offer payback periods of around 10 years versus oil for boilers of up to 14 MW, coal is clearly much more attractive economically.

'From Fig. 6.14 it can also be seen that a farm with insufficient of its own straw (or a horticultural enterprise importing straw) and buying straw for a boiler at £22/tonne achieves rather poorer payback periods versus coal and oil firing than a farm with its own straw.'

Plate 6.6 A relatively simple triple pass system gets a useful increase in efficiency over the original single pass, unsophisticated burners which were really just incinerators.

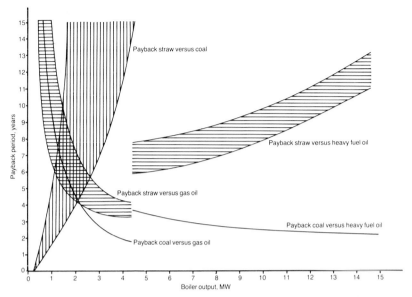

Fig. 6.14 Economics of straw fired boilers vs. coal and oil fired boilers for off-farm applications.

ACTION PLAN — FUEL

Annual fuel bill

If fuel bill is in £1000 – £3000 range and labour available and some flue gases acceptable → Use conventional bales in single or triple pass boiler — hand fed.

If possible combination of fuel bills over £4000 – £5000 *or* if fuel gases need to be clean → Use high efficiency furnace and automatic feed with conventional bales. Possibly consider extra cost of large bale feed system if labour limiting.

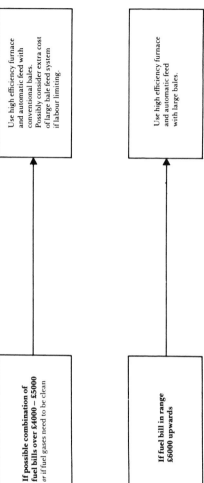

If fuel bill in range £6000 upwards → Use high efficiency furnace and automatic feed with large bales.

Index

Bold type signifies major sections or chapters